RESIDUE REVIEWS

VOLUME 29

SPECIAL VOLUME—SYMPOSIUM ON

DECONTAMINATION OF PESTICIDE RESIDUES IN THE ENVIRONMENT

ATLANTIC CITY MEETINGS
OF THE
AMERICAN CHEMICAL SOCIETY
September 1968

SPRINGER-VERLAG
BERLIN · HEIDELBERG · NEW YORK
1969

RESIDUE REVIEWS

Residues of Pesticides and Other
Foreign Chemicals in Foods and Feeds

RÜCKSTANDS-BERICHTE

Rückstände von Pesticiden und anderen
Fremdstoffen in Nahrungs- und Futtermitteln

Edited by

FRANCIS A. GUNTHER

Riverside, California

VOLUME 29

SPRINGER-VERLAG
BERLIN · HEIDELBERG · NEW YORK
1969

ISBN 978-1-4615-8457-5   ISBN 978-1-4615-8455-1 (eBook)
DOI 10.1007/978-1-4615-8455-1

© 1969 by Springer-Verlag New York Inc.
Softcover reprint of the hardcover 1st edition 1969
Library of Congress Catalog Card Number 62–18595.

Title No. 6632

Preface

That residues of pesticide and other "foreign" chemicals in foodstuffs are of concern to everyone everywhere is amply attested by the reception accorded previous volumes of "Residue Reviews" and by the gratifying enthusiasm, sincerity, and efforts shown by all the individuals from whom manuscripts have been solicited. Despite much propaganda to the contrary, there can never be any serious question that pest-control chemicals and food-additive chemicals are essential to adequate food production, manufacture, marketing, and storage, yet without continuing surveillance and intelligent control some of those that persist in our foodstuffs could at times conceivably endanger the public health. Ensuring safety-in-use of these many chemicals is a dynamic challenge, for established ones are continually being displaced by newly developed ones more acceptable to food technologists, pharmacologists, toxicologists, and changing pest-control requirements in progressive food-producing economies.

These matters are of genuine concern to increasing numbers of governmental agencies and legislative bodies around the world, for some of these chemicals have resulted in a few mishaps from improper use. Adequate safety-in-use evaluations of any of these chemicals persisting into our foodstuffs are not simple matters, and they incorporate the considered judgments of many individuals highly trained in a variety of complex biological, chemical, food technological, medical, pharmacological, and toxicological disciplines.

It is hoped that "Residue Reviews" will continue to serve as an integrating factor both in focusing attention upon those many residue matters requiring further attention and in collating for variously trained readers present knowledge in specific important areas of residue and related endeavors; no other single publication attempts to serve these broad purposes. The contents of this and previous volumes of "Residue Reviews" illustrate these objectives. Since manuscripts are published in the order in which they are received in final form, it may seem that some important aspects of residue analytical chemistry, biochemistry, human and animal medicine, legislation, pharmacology, physiology, regulation, and toxicology are being neglected; to the contrary, these apparent omissions are recognized, and some pertinent manuscripts are in preparation. However, the field is so large and the interests in it are so varied that the editor and the Advisory Board earnestly solicit suggestions of topics and authors to help make this international book-series even more useful and informative.

"Residue Reviews" attempts to provide concise, critical reviews of timely advances, philosophy, and significant areas of accomplished or needed endeavor in the total field of residues of these chemicals in foods, in feeds, and in transformed food products. These reviews are either general or specific, but properly they may lie in the domains of analytical chemistry and its methodology, biochemistry, human and animal medicine, legislation, pharmacology, physiology, regulation, and toxicology; certain affairs in the realm of food technology concerned specifically with pesticide and other food-additive problems are also appropriate subject matter. The justification for the preparation of any review for this book-series is that it deals with some aspect of the many real problems arising from the presence of residues of "foreign" chemicals in foodstuffs. Thus, manuscripts may encompass those matters, in any country, which are involved in allowing pesticide and other plant-protecting chemicals to be used safely in producing, storing, and shipping crops. Added plant or animal pest-control chemicals or their metabolites that may persist into meat and other edible animal products (milk and milk products, eggs, etc.) are also residues and are within this scope. The so-called food additives (substances deliberately added to foods for flavor, odor, appearance, etc., as well as those inadvertently added during manufacture, packaging, distribution, storage, etc.) are also considered suitable review material.

Manuscripts are normally contributed by invitation, and may be in English, French, or German. Preliminary communication with the editor is necessary before volunteered reviews are submitted in manuscript form.

Department of Entomology   F.A.G.
University of California
Riverside, California
September 8, 1969

Table of Contents

Introduction to the symposium:* Decontamination of pesticide residues in the environment

By

G. K. KOHN **

This division of the American Chemical Society is concerned with food and with agriculture. The members of this division are all involved in research and development aimed at improvement of the world food supply both quantitatively and qualitatively, and the development of the art and science of agriculture.

The purposes of this symposium are:

1. Demonstrate the concern by the professional chemist for the problem of pesticide residues.
2. Remove the discussion from the realms of emotion and polemic (for all sides of the issue) to the cooler climate of logic and scientific investigation.
3. Review the present status of programs for pesticide decontamination in the United States and in the rest of the world.
4. Explore current needs and future programs and encourage greater research effort for the future.

One of the tools that has contributed much to increased food availability has been the employment of inorganic and organic chemicals which effect the reduction or elimination of undesirable species (plants, insects, rodents, fungi, bacteria, virus, etc.) and the nutrition, growth, and reproduction of the desirable species. The use of chemicals to these ends contributes to ecological change. It is by now an accepted commonplace, but it requires reiteration that all of man's efforts to improve and increase his food supply, including cultivation and primitive agriculture themselves, significantly altered the environment. That, indeed, was their objective. As chemists and as responsible citizens, it behooves us from time to time to review the broader implications of our investiga-

* Presented September 9 and 10, 1968 during the 156th National Meeting, American Chemical Society, Atlantic City, New Jersey.
** Chevron Chemical Company, Ortho Division, Richmond, California.

1

tions where ecology has been upset. We hope to communicate to those concerned with agriculture and to the public at large the results of such review.

The scientist in all of us knows that as we apply the results of our creative efforts and alter the environment to derive beneficial results particularly, but not exclusively, in biologically related fields, there are reactions and interactions that must result.

Elsewhere you can find the statistics that attest to the tremendous achievements in the productivity of American agriculture in great part the result of the application of chemistry and mechanics to this most fundamental of man's activities. You will find elsewhere also a growing concern for the ecological consequences of this achievement. It is the conviction of the chairman of this symposium that the world requires more than ever broader and broader applications of science and the scientific method. While it is true that each new application of the results of scientific investigation may create a new hazard and a new problem, it is equally true that the cessation of scientific effort and its application results in a sterile inaction and in ultimate disaster.

The following papers discuss within the framework of this philosophy programs and methods for the reduction of pesticide residues after their application. In doing so, we recognize that certain chemicals are useful in agriculture and for society. At the same time, there is a growing need to control their residues, to minimize harmful ecological consequences, and to weigh both the achievements in productivity and these ecological consequences therefrom on the balances of scientific logic. In this spirit there follow papers on a variety of topics aimed at the reduction of the hazards relating to pesticide residues. Among them will be papers on the methods of food processing and their affects on residue reduction or elimination, on reduction of residues in water supplies, on treatment of soil and its effect on residues, on removal of residues from animals feeds, and on the status and programs for residue reduction in countries other than our own.

Attenuation of pesticidal residues on seeds

By

G. K. KOHN [*]

Contents

I. Introduction

The farmer, from the beginning of pre-history to the present, has been confounded with the hazards associated with preservation of seeds, their planting, their germination, and the nurture of the delicate seedling that might possibly emerge. Modern scientific agriculture has eliminated much of this hazard by treating this seed with a variety of chemical substances. These provide protection from insects, from rodents, from birds, and from diseases. In some cases these chemicals protect against excessive heat and cold. They stimulate growth and provide nutrition. These seed treatments for highly advanced agriculture, as in our country, are frequently performed by large seed producing companies that not only through hybridization produce the optimum seed variety for a given geographical region but perform the treating operations. These treatments are also performed by other associations and individuals. Because at this period one cannot predict weather, infestation, disease, or epidemic conditions with certainty each year much more seed has to be treated than is used. Despite our relative sophistication, second and third plantings are common. It, therefore,

---

[*] Chevron Chemical Company, Ortho Division, Richmond, California.

becomes necessary each year to treat much more seed than is actually planted.

Although the pesticides [1] and other chemicals used in these treatments generally leave no harmful residues in the later harvested crop, the treatments used for seed protection do leave high residues on the treated seeds. In Table I we have the expected normal residue from a commercial seed treatment of a single pesticide. Frequently these residues can be higher for multiple treatment as, for example, when an insecticide and a fungicide are applied to the same seed.

Table I. *Approximate residues on treated seed*

Active ingredients applied (oz./cwt.)	Range of expected residue (p.p.m.)
0.25	100- 200
0.5	200- 400
1.0	400- 800
2.0	800-1,600

These will vary, of course. One never gets 100 percent efficient transfer. There are flaking and surface removal, uneven application, and losses from volatilization and from chemical interaction, etc. The magnitudes of the residues are high and generally far in excess of legal tolerances for human consumption where they exist or would be likely to exist in the future for any given chemical.

Some of the principal crops where seed treatments are significantly prevalent are presented in Table II.

Table II. *Seed treatment for major crops in the United States*
1966-1967

Crop	10^6 Acres planted [a]	Approximate lb. seed/acre	Seed treated (%)	10^6 Lb. treated
Corn	66.3	8.0	100	530.4
Wheat	54.5	60	75-80	253.0
Sorghum	16.3	25	100	40.8
Cotton	10.3	33.3	90-100	343.0
Edible beans	1.9	50	100	95.0
Peanuts	1.5	33.3	100	50.0

[a] Agricultural statistics, U. S. Government Printing Office, Washington, D. C. (1966 and 1967).

[1] Common and chemical names of pesticides mentioned in text are listed in Table X.

From Table I it is evident to those familiar with tolerance ranges for common pesticides that the magnitude of the residues found would prohibit in almost all cases the use of these seeds for human and from animal consumption. To protect the public against error and possible unscrupulous practice as in adulteration of normal cereal products all seed treated with pesticides must by law incorporate a dye in the formulation (*Federal Register* 1962). Our objective in the experiments to be described was to reduce the residues materially (100- to 1,000-fold) and, if possible, completely, so that within the limitations of FDA tolerances certain pesticide-treated seed could possibly be utilized for animal feeds and less probably even for human food. Protection of the public would result, and economies for the seed companies, and particularly for the farmer, would accrue.

II. Chemical considerations

Within the limitations of the economics of the seed treating industry and the need for the use of "safe" reagents only the simplest of chemical and physical treatments would be permitted. Our experiences in the areas of the metabolic fate of certain pesticides lead us to the consideration of nucleophilic displacement reactions for the rapid and efficient destruction of certain groups of pesticides. Among the nucleophiles examined we chose primarily (but not exclusively) sulfur-containing anions, a lead that derived from tables published by STREITWEISER (1962) on relative reaction rates.

From work on the metabolic fate of fungicides such as CAPTAN, PHALTAN® and DIFOLATAN® [2] it was found that the scission of the N-S bond in homogeneous aqueous solutions treated with RS⁻ is quite rapid and is followed by the destruction of the resulting sulfenic acid to give various sulfur species and ultimately quantitative chloride ion accountability, as in Figure 1:

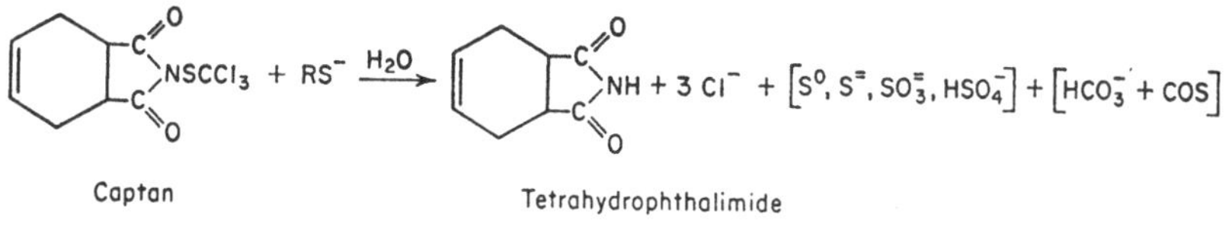

Fig. 1. Captan degradation by reaction with nucleophiles (condensed)

For a pesticide such as DIBROM® the first homogeneous reaction is extraordinarily rapid and grossly almost instantaneous.[2] In this case we write only the first reaction, as in Figure 2:

---

[2] Unpublished data submitted in documents by the *Chevron Chemical Company* to the *U. S. Department of Agriculture* and to the *Food and Drug Administration* in various registration applications.

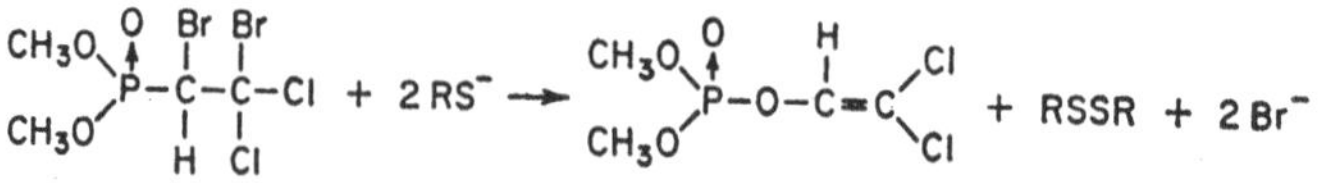

Dibrom DDVP

Fig. 2. Dibrom decomposition: when RS^- is cysteine ion, RSSR is cystine

Another type of displacement reaction occurs with useful rapidity with methyl phosphate insecticides. It consists of the demethylation reaction [3] and we illustrate this reaction with the insecticide PHOSPHAMIDON as in Figure 3:

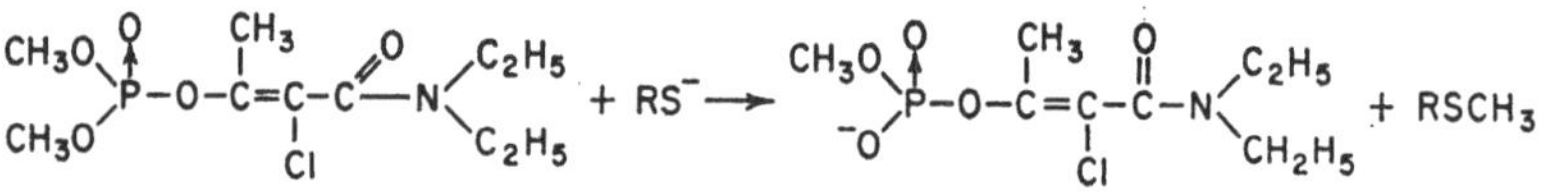

Phosphamidon Desmethylphosphamidon

Fig. 3. Phosphamidon interaction with RS^-

The above reactions are illustrations only. Actually several concurrent and consecutive reactions may take place and they will differ with each insecticide. In general we have shown that pesticides having labile halogens (that is aliphatic halogens activated by proximity to sulfur or oxygen as in carbonyl group, or sulfenic linkage, etc.) or aromatic halogens *ortho* or *para* to strong electron-withdrawing groups are amenable to such displacements. Other pesticides such as the methyl phosphate esters are also attacked by the nucleophiles. Examples are METHYL PARATHION, BIDRIN,® etc. Certain carbamates and nonmethyl phosphates having groupings amenable to such attack are, of course, candidates for this type of treatment. The carbamate linkage is frequently susceptible to such degradation. Excluded as candidates are systemic compounds and stable halogen linkages as in DDT (we will illustrate such failures with DDT and lindane). The most favored compounds are described at length in a recently issued United States patent (1968).

III. Experimental procedure and results

a) Model trials–formulations

Studied first were formulations of dusts used commonly for seed

[3] Based on an assay method for Phosphamidon originally developed in the *Ciba Analytical Laboratories*, Basle, Switzerland.

treatment but without the coating of the seed. The most convenient laboratory method employed was to add given quantities of pesticide formulation (and later treated seed) to a flask to which could be added the required nucleophile with or without buffering chemicals and surfactants. In the case of the model experiments with formulations the pesticide was analyzed by standard published methods usually, by preference, vapor phase chromatography (VPC) or quantitative assays from thin-layer plates. A powerstat was employed to govern the rate of rotation of a rotating flash evaporator. Since minimum damage to the seed was an objective, this type of gentle rotation appeared to be well suited as a technique. Records of temperature, pH (before and after), and concentration of chemical reagents enabled us to standardize the procedure and furnished a basis for later larger scale extrapolation.

Table III provides results of experiments with lindane and DDT and illustrates examples of those halogenated pesticides where the chemical degradation is slow or negligible. Some of the cyclodiene insecticides would be in this category for certain of them are too stable to show appreciable degradation. As might be expected DDT gave

Table III. *Chemical stability of certain organochlorine compounds*

Pesticide	Nucleophile	Degradation (%)
Lindane	$CaS_{4.5}$	11 (50°C., 1 hr)
DDT	$CaS_{4.5}$	< 1 (25°C., 24 hr.)

practically no breakdown under these conditions. On the other hand, a formulation of malathion when similarly treated illustrated the relative effects of RS^- as compared to OH^- as nucleophiles. In this experiment 100 mg. of malathion 25 percent wettable powder was added to 40 ml. of water with one, two, and three ml. of the previously employed ORTHORIX® SPRAY [4] as nucleophilic reagent. The flasks were extracted with benzene and the solution was assayed by TLC (Table IV).

A typical halogenated pesticide that was attacked with reasonable rapidity is provided in the next experiment. Exactly 250-mg. portions of an 80 percent DIFOLATAN® formulation was suspended in 500 ml. of water buffered to the desired pH at ambient temperature. Chloride ion titration was employed to measure destruction of DIFOLATAN®. Exactly 1.5 ml. of mercaptoacetic acid was the nucleophile (Table V).

---

[4] A lime sulfur composition containing 25 percent calcium polysulfide and 10 percent non-ionic surfactant.

Table IV. Destruction of malathion at 25°C. by calcium polysulfide

Rotation time (min.)	Malathion remaining, polysulfide treatment (%)		
	1 ml.	3 ml.	40 ml.
5	55	40	30
20	30	13	8.5
40	5	0.4	nd [a]
80	nd	nd [a]	nd [a]

[a] nd = below limit of assay method (0.4 percent).

Table V. *Destruction of DIFOLATAN 80W by mercaptoacetic acid*

Time in flask (hr.)	Destroyed (%)
0	0
4	90
24	99.3
4 [a]	1.8

[a] No mercaptoacetic acid.

b) Treated seed experiments

The attenuation of residues from treated seeds is both a physical and a chemical process and these processes are not entirely independent of each other. A residue on the treated seed can be partially washed or rubbed off by physical means. Where a chemical reaction destroys part of the residue on the seed the physical removal is aided and it in turn provides fresh surface for chemical reaction.

The methods used for analysis were very similar. In these cases a sample of the treated seed would be appropriately extracted prior to the treatment and the extract analyzed by conventional means. Samples would be taken from the treatments at various intervals and similarly extracted and analyzed.

In Table VI is illustrated the rapid degradation of residues of DIFOLATAN® wettable at a rate of one oz./bushel of seed. Fifty g. of treated seed was placed in the rotating flask containing three ml. of the ORTHORIX® formulation and enough water added to cover the seeds. Analysis included extractions and VPC assay.

Similar experiments are summarized in Table VII for cotton seed (V. Acala 4-42) treated with DIFOLATAN® seed transfer formulation,

Table VI. *Removal of DIFOLATAN residues from hybrid seed corn at 25°C.*

Rotation time (min.)	Residue (p.p.m.)
0	626
10	2.25
60	0.2

Twenty-five g. of seed were rotated 10 minutes with varying quantities of polysulfide solutions. Similarly, the data in Table VIII are for some varieties of cotton seed treated with Captan. Other experiments were

Table VII. *Removal of DIFOLATAN residues from cotton seeds*

Delinting process	Polysulfide added (ml.)	Treatment time (min.)	Residue (p.p.m.)
Acid	0	0	464
	1	10	3.2
	3	10	2.8
	5	10	0.4
Mechanical	0	0	702
	1	10	14
	3	10	6.6
	5	10	4

Table VIII. *Removal of CAPTAN residues from cotton seeds*

Delinting process	Polysulfide added (ml.)	Treatment time (min.)	Residue (p.p.m.)
	0	0	750
Acid	1	10	1.3
	3	10	0.8
Mechanical	0	0	1124
	1	10	4
	3	10	1.6
	5	10	0.5

run with sodium sulfide, ammonium polysulfide, sodium sulfite, sodium thiosulfate, various mercaptides, etc. Economics highly favor simple inorganic salts. Agitation rates studies are indicative but must be balanced against the practical problems of commercial processing and of seed damage.

Corn seed treated with DIFOLATAN® was run with varying rotation speeds for a treatment time of three minutes using 25 g. of seed in 40 ml. of water and 0.3 ml. of polysulfide solution (Table IX). Some

Table IX. *Effects of agitation rate on residues of DIFOLATAN on corn seed at 30°C.*

Agitation (3 min.)	Residue after treatment (p.p.m.)
Mild	11.0
Moderate	6.0
Vigorous	1.5
No treatment	485

pilot attempts have been initiated for scaling up this work. A screw conveyor carries the seed through the treating solution where it can be washed and dried.

IV. Limitations and extensions

The treatment we have described of course does not extend to systemic chemicals or chemicals readily solubilized by the natural seed coat. Treatments that would affect such residues would generally deteriorate the seed itself. Pesticides that are more or less insoluble and remain to a large degree on the surface are, of course, preferred candidates for this treatment. As we have previously discussed, the pesticide must be amenable to nucleophilic displacement reactions. Surfaces other than seeds that possess pesticidal residues can be similarly treated and are subject to the same limitations and may provide particular additional limitations. We believe that, within the limitations cited, important crop seeds can be treated as described with reduction of residues from the hundreds or thousands of p.p.m. to a range of between 0.1 to 10 p.p.m., depending upon the initial mode and time of pesticide treatment and the reaction conditions employed, both chemical and physical. This method provides, then, a method for the decontamination of a small part of man's environment. The procedure can be extended and could reduce the hazard associated with pesticides residues entering primarily the animal food chain and secondarily man's.

Summary

The need for and the extent to which seeds are treated with pesticidal and fertilizer chemicals are briefly discussed as are the quantity of normal residues on the seed resulting from such treatment and the possible hazards therefrom. Certain pesticidal chemicals are subject to relatively easy destruction by nucleophilic reagents. Among these are the sulfene imide fungicides, certain esters of phosphoric acid, particularly methyl esters. Washing treatments with inorganic sulfides,

Table X. *Common and chemical names of pesticides used in this investigation*

Naled (DIBROM®)	1,2-Dibromo-2,2-dichloroethyl dimethyl phosphate
Captan (ORTHOCIDE®)	N-Trichloromethylthio-4-cyclohexene-1,2-dicarboximide
Folpet (PHALTAN®)	N-Trichloromethylthiophthalimide
DIFOLATAN®	N-(1,1,2,2-Tetrachloroethylthio)-4-cyclohexene-1,2-dicarboximide
Malathion	S-(1,2-Dicarbethoxyethyl) O,O-dimethyl phosphorodithioate
Lindane	Gamma isomer of 1,2,3,4,5,6-hexachlorocyclohexane
DDT	2,2-Bis(p-chlorphenyl)-1,1,1-trichloroethane
Methyl Parathion	O,O-Dimethyl O-p-nitrophenyl phosphorothioate
Phosphamidon (DIMECRON®)	1-Chloro-3-(dimethoxyphosphinyloxy)-N,N-diethyl-cis-crotonamide
Bidrin®	3-(Dimethoxyphosphinyloxy)-N,N-dimethyl-cis-crotonamide
Ciodrin®	alpha-Methylbenzyl 3-(dimethoxyphosphinyloxy)-cis-crotonate

polysulfides, organic thiols, etc. can reduce seed residues from 500-1,000 ppm to one p.p.m. and less depending upon the vigor and extent of treatment.

Résumé *

Diminution des résidus de pesticides sur les semences

La nécessité de traiter les semences par des pesticides et des engrais chimiques et l'importance de ces traitements sont brièvement discutés, ainsi que les résidus qui en résultent normalement et les dangers potentiels qu'ils entraînent. Certains pesticides chimiques font l'objet d'une destruction relativement aisée par des réactifs nucléophiles. Parmi ceux-ci figurent les fongicides du groupe des sulfène-imides, certains esters de l'acide phosphorique, en particulier, les esters méthyliques. Des traitements par lavage à l'aide de sulfures inorganiques, de polysulfures, de thiols organiques, etc. peuvent réduire les résidus sur les semences de 500 à 1000 p.p.m. jusqu'à 1 p.p.m. ou moins, selon l'intensité et l'importance du traitement.

Zusammenfassung **

Verkleinerung von Pestizidrückständen auf Samen

Die Notwendigkeit und das Ausmass, Samen mit Pestiziden und

---

* Traduit par S. DORMAL-VAN DEN BRUEL.
** Übersetzt von A. SCHUMANN.

Düngungschemikalien zu behandeln, werden kurz diskutiert, sowie auch die normalen Rückstandsmengen auf Samen, die von solchen Behandlungen herrühren, und die sich daraus ergebenen möglichen Gefahren. Gewisse Pestizidchemikalien lassen sich relativ leicht mit nukleophilen Reagentien zerstören. Zu diesen gehören die Sulfimid- fungizide, einige Ester der Phosphorsäure, insbesondere Methylester. Waschbehandlungen mit anorganischen Sulfiden, Polysulfiden, organi- schen Thiolen usw. können, je nach Stärke und Ausmass der Behand- lung, Samenrückstände von 500-1000 ppm zu 1 ppm und weniger reduzieren.

References

Federal Register (27 F.R. 10494), Title 21, Food and Drug, Chapter 1, Sub-
    chapter A, parts 3, 3.13 (a and b). Oct. 27, 1962.
Streitweiser, A.: Solvolytic displacement reactions at saturated carbon atoms. Chem.
    Reviews **56**, 583 (1962).
U.S. Patent no. 3,305,441 (1968).

The decontamination of animal feeds

By

T. E. ARCHER * AND D. G. CROSBY *

Contents

I. Introduction

Pesticides have become essential to the economical production of most animal feedstuffs, including also the by-products of other agricultural industries, for example almond hulls, seed screenings, fruit pomaces, and other materials. The treatment which results in contamination may be intentional such as acaricide applications to almond hulls, or incidental such as drift onto alfalfa fields, but an increasing proportion of feed samples contain unacceptable levels of pesticide residues.

The Commercial Feed Regulations of the *California Department of*

* Department of Environmental Toxicology, University of California, Davis, California.

13

Agriculture issued in 1968 define a "commercial" feed as including all substances which are used as feed or for mixing in feed for animals except as specifically exempted in Section 14902. This regulation also defines the permitted tolerances for the following pesticides on animal feeds in California: 1) DDT, DDD, DDE total residue: 0.5 p.p.m. in or on commercial feed for lactating and nonlactating dairy animals; 2) Kelthane[1,1-bis(p-chlorophenyl)-2,2,2-trichlorethanol]; 10 p.p.m. in or on any ingredient to be used in the manufacture of a complete commercial feed ration, provided the complete commercial feed ration contains not more than 1.5 p.p.m. and is sold only for use as a feed for ruminant meat animals. The permissible tolerance for dairy animals and poultry is still zero p.p.m.

Table I contains examples of a few of the major animal feeds

Table I. *Examples of pesticides found by the California Department of Agriculture in commercial animal feeds in 1967*

Material	Production as feed (approx. tons/yr.)	Samples with overtolerance residues (approx. %)	Type of contaminant as residue
Alfalfa products	5,000,000	5	DDT, toxaphene
Almond hull meal	100,000	34	Kelthane, DDT, DDD
Apple pulp	30,000	93	Kelthane, DDT, DDD
Seed screenings	20,000	42	Toxaphene, DDT, DDD, Kelthane
Grape pomace	50,000	47	Kelthane, DDT
Raisin stem meal	20,000	80	Kelthane

analyzed by the *California Department of Agriculture* Feed and Livestock Remedy Laboratory in 1967 for pesticide contamination. The principle offending pesticides in the feeds analyzed were DDT and its analogs, toxaphene, and Kelthane. Since the percentages of samples containing over tolerance pesticide contaminants ranged from five to 93 percent, these data emphasize the need for the decontamination of animal feeds.

Organochlorine insecticides have been predominantly responsible for feed contamination. Removal of these residues may be accomplished by exploitation of some characteristic physical or chemical property such as volatility, instability to hydrolysis, or other treatments. Practical scale examples of the removal of specific pesticide residues indicates that decontamination methods demonstrated in the laboratory may become economically feasible.

The purpose of the present work was to find means by which pesticide residues on common animal feeds might intentionally be removed or detoxified. Decontamination methods would improve both the value of the feed as a marketable commodity and the potential marketability of animal products containing residues derived from the feed source; they would be of particular value in instances of accidental over-application or drift contamination. Observations that DDT is relatively volatile, especially in the presence of water (BOWMAN *et al.* 1959) and remains near the plant surfaces in alfalfa (ARCHER and CROSBY 1967), might provide the basis for effective practical removal of pesticide residues from animal feedstuff.

II. Methods and materials

a) Analytical methods: Methods for the detection and determination of the pesticides

Thin-layer chromatography (TLC) and gas-liquid chromatography (GLC) procedures were employed routinely, either separately or in combination. The GLC column was eight feet of ⅛-inch O.D. stainless steel packed with 60 to 80 mesh silylated chrom W coated with five percent SE 30 plus five percent Dow 710 silicone oil. The column oven temperature was 200°C.

TLC was employed for screening using a 0.5-mm. layer thickness of Silica Gel H and, in combination with GLC, as an analytical tool. Developing solutions, chromogenic reagents, and R_f values are explained in Table II. For quantitative work, chromatogram areas containing the unknowns were extracted from the silica gel with benzene or pentane after comparing R_f values of the plant extracts with those of parallel standard pesticide tracers, and the extracts were analyzed by GLC.

b) Sample preparation

1. **Hay samples.** — Hay samples containing DDT and its analogs were obtained from fields in Yolo County, California. The fields had not received intentional treatment with pesticides during the previous season, so the existing residues represent accidental contamination. Moisture content was determined at the time of treatment (immediately after harvest in the case of fresh alfalfa): field-cured hay contained 16.5 percent, air-dried green-chop 9.5 percent, commercially dehydrated meal 5.0 percent, and alfalfa pellets 5.8 percent moisture. Chopped, mixed, composited hay samples were subsampled for use in the various experiments. The hay sample for the endrin experiments was prepared in the following manner. A previously analyzed field-cured hay sample (DDT residue, 0.05 p.p.m., nine percent moisture) was cut into three to six mm. pieces and mixed. A weighed amount of

Table II. *Thin-layer chromatography separation of DDT and analogs, endrin, Kelthane, and 4,4-dichlorobenzophenone on Silica Gel H*

Pesticide	Developing solvents	Chromogenic reagent mixture	R_f values
DDD	100% Pentane	Silver nitrate,[a] 2-phenoxyethanol	0.14
DDT	100% Pentane	Silver nitrate,[a] 2-phenoxyethanol	0.26
DDE	100% Pentane	Silver nitrate,[a] 2-phenoxyethanol	0.41
Endrin	100% Benzene	Indophenol blue,[b] silver nitrate, citric acid	0.76
Kelthane	5% Isopropyl alcohol, 95% pentane	Silver nitrate,[a] 2-phenoxyethanol	0.50
4,4'-Dichlorobenzo- phenone	5% Isopropyl alcohol, 95% pentane	Silver nitrate,[a] 2-phenoxyethanol	0.67
Malathion	2% Isopropyl alcohol 98% benzene	Tetrabromo-[c] phenolphthalein ethyl ester, silver nitrate, citric acid	0.50

[a] MITCHELL (1958).
[b] GRAHAM (1963).
[c] KOVACS (1964).

the mixed hay was covered with redistilled acetone containing a known amount of analytical grade endrin to result in a theoretical residue deposit of 1,000 p.p.b. on the hay. The solvent was evaporated at 50 to 60°C. with the aid of an air stream until completely dry. The hay was mixed in the solvent concentrating jar and stored for analysis. By analysis, it contained a residue of 900 p.p.b. of endrin.

2. Dry almond hull meal and ladino clover seed crop screenings samples. — Previously analyzed dry, ground almond hulls (14 percent moisture) and ladino clover seed screenings (13.3 percent moisture) were well mixed separately. A weighed amount of the mixed hulls was covered with redistilled acetone containing a known amount of analytical grade Kelthane to result in a theoretical residue deposit of 15,000 p.p.b. on the hulls. A weighed amount of the mixed seed crop screenings was covered with redistilled acetone containing a known amount of analytical grade malathion to result in a theoretical residue deposit of 60,000 p.p.b. on the screenings. The solvent was evaporated from both samples at 50 to 60°C. with the aid of an air stream until completely dry. The plant materials were separately mixed in the solvent concentrating jars and stored for analysis. By analysis the almond hulls contained 10,200 p.p.b. Kelthane and the seed crop screenings contained 40,000 p.p.b. of malathion after coating.

air-dried at room temperature for five days in a hood (10 percent moisture), commercially dehydrated meal obtained from the green-chop

Table III. *Vigorous water wash to determine if pesticide residues [a] on alfalfa plant material were a loose surface deposit*

Sample	DDT-RCH [b] (p.p.b.)		Endrin (p.p.b.)	
	Cold water treatment	Hot water treatment	Cold water treatment	Hot water treatment
Content before treatment	560	560	900	900
Content after treatment	544	384	928	835
Loss due to treatment	2.9%	31.4%	±	7.2%
Content of water wash #1	46.3	8.8	38.0	73.0
Content of water wash #2	5.5	5.0	26.0	36.0
Content of water wash #3	10.5	2.6	18.0	16.0
Total residue reclaimed	108%	71.4%	112%	107%

[a] Residues calculated on a dry weight basis.
[b] RCH = related chlorinated hydrocarbons.

c) Pesticide extraction

Extraction of the plant samples (15 g./250 ml. of solvent) was accomplished by three ½-hour refluxes with benzene; the solvent was pooled, concentrated, and analyzed. The solvent extracts were cleaned-up on florisil activated at 270°C. for three hours. DDT and its analogs, endrin, Kelthane, and 4,4'-dichlorobenzophenone were eluted from the florisil with 390 ml. of 30 percent diethyl ether and 70 percent pentane. Malathion and malaoxon were eluted from the florisil with 390 ml. of two percent isopropyl alcohol and 98 percent benzene.

d) Dehydration

Alfalfa was dehydrated rapidly at inlet temperatures of 760° to 1000°C. and an outlet temperature of 127°C. for three to five minutes in a revolving drum type dehydrator. Two to three tons of green alfalfa/ hour were dehydrated under these conditions. Four types of samples were analyzed: green-chop containing 84 percent moisture, green-chop

Table IV. *Vigorous water wash* [a] *to determine if pesticide residues on almond hulls were a loose surface deposit*

Sample	Kelthane (p.p.b.)		4,4'-D [b] (p.p.b.)	
	Cold water [c] treatment	Hot water [d] treatment	Cold water treatment	Hot water treatment
Residue [e] before treatment	10,200	10,200	N.D.[f]	N.D.[f]
Residue after treatment	6,500	4,800	1,300	1,300
Residue in combined water washes	N.D.[f]	N.D.[f]	N.D.[f]	N.D.[f]
Kelthane residue loss due to treatment	36%	53%	—	—
Total residue loss due to treatment, calc. as Kelthane	19%	35%	—	—
Total residue recovered, calc. as Kelthane	81%	65%	—	—

[a] Washed three times for 15 minutes with vigorous agitation.
[c] 4,4'-D signifies 4,4'-dichlorobenzophenone.
[c] Cold water temperature 25° C.
[d] Hot water temperature 90° to 100° C.
[e] All residues calculated on a dry weight basis.
[f] N.D. = non-detectable.

(five percent moisture), and pelleted hay obtained from the same green-chop (six percent moisture). The green-chop represented composited samplings of several acres of harvested alfalfa.

III. Results and discussion

a) Water washing

The vigorous water washing experiments were performed by placing a weighed amount of plant material in a large Erlenmeyer flask with 30 volumes of either cold (23°C.) or hot (90° to 100°C.) distilled water and shaking vigorously for 15 minutes on a rotary laboratory shaker. The water was filtered through a small, loosely packed glass wool plug and stored. Washing was performed three times; the glass wool plug was extracted with the first water wash three times with benzene and

Table V. *Vigorous water wash* [a] *to determine if pesticide residues on ladino clover seed crop screenings were a loose surface deposit*

Sample	Malathion (p.p.b.)		Malaoxon (p.p.b.)	
	Cold water [b] treatment	Hot water [c] treatment	Cold water treatment	Hot water treatment
Residue [d] before treatment	40,000	40,000	N.D.[e]	N.D.[e]
Residue after treatment	5,700	4,000	N.D.[e]	N.D.[e]
Residue in combined water washes	3,100	4,200	N.D.[e]	N.D.[e]
Malathion residue loss due to treatment	86%	90%	—	—
Total residue recovered calc. as malathion	22%	21%	—	—

[a] Washed three times for 15 minutes with vigorous agitation.
[b] Cold water temperature 25° C.
[c] Hot water temperature 90° to 100° C.
[d] All residues calculated on a dry weight basis.
[e] N.D. = non-detectable.

the contents were pooled and analyzed. The second and third water washes were extracted in a similar manner without the glass wool plug.

The data in Table III show the nature of the distribution of the residues on hay. After three vigorous washes with either hot or cold water, very little DDT-RCH (DDT and related chlorinated hydrocarbon) and endrin residue was removed from the plant material, and only traces of contaminants were found in the washes. This indicates that the contaminants were not loosely distributed on the surface of the plant material as a thin film or deposited on the hay in dust or soil. The higher loss of residue with the hot water treatment may be explained by the volatility of DDT and endrin with water vapors. Others (ACREE *et al.* 1963) have observed that DDT was lost from various media as a result of the DDT-water codistillation phenomenon.

The data in Table IV show that very little Kelthane or 4,4′ D (4,4′-dichlorobenzophenone) was removed from almond hulls by three vigorous washes with either cold or hot water. No residues could be detected in the water washes, but 36 percent of the Kelthane residues

was removed from the hulls by the cold water wash and 53 percent
was removed by the hot water wash. If the 4,4'-D residues found on
the hulls after treatment were converted to Kelthane and summed with
the Kelthane residues found after treatment, 19 percent of the original
Kelthane residue was removed by the cold water wash and 35 percent
was removed by the hot water wash. If the total residue found after
treatment were calculated as Kelthane, 81 percent of the original residue was recovered for the cold water wash and 65 percent for the
hot water wash. The higher loss of residue with the hot water treatment may possibly be explained as volatility of the residue with
water vapors.

Table V contains data for the removal of malathion from ladino
clover seed crop screenings by vigorous hot and cold water washings.
Both hot and cold water were effective in the removal of the malathion
residues: 86 percent removal for the cold water treatment and 90
percent for the hot water treatment. No malaoxon was detected in the
samples analyzed; only traces of malathion could be detected in the
water washings.

b) Solvent washing

If the residue were not loosely deposited on the alfalfa hay, then
it might be assumed that the DDT and endrin residues may be deposited in such sites as the wax-like materials of the plant cuticle. The
experiment designed to remove the pesticide residue and cuticle waxes
was accomplished by washing 60 g. of composited sample for three
minutes with 900 ml. of benzene warmed to 80°C., filtering the solvent
through a single layer of cheesecloth, concentrating the solvent, clean-
ing-up the sample on florisil, and analyzing it. The contaminated hay
was washed with benzene seven times, and the remaining plant material was extracted by three benzene refluxes.

Table VI shows that after seven benzene washes at 80°C., 96 percent of the DDT-RCH residue and 95 percent of the endrin residue
were removed. The original residues present on the hay were essentially recovered in the benzene washes, indicating a concentration of
the contaminants in the plant cuticle.

c) Warm air treatment

The effectiveness of oven heat at 100°C. for the removal of the
pesticide contaminants from the plant materials was investigated.
A warm air, convection-type oven controlled at $100° + 5°C$. was used
for the oven heating experiments. The dry heating experimental treatment was for 12 hours. The wet heating experiment involved saturating
a weighed amount of mixed and composited plant material with water,

Table VI. *DDT-RCH and endrin residues* [a] *in alfalfa plant material to determine if the pesticides were deposited in sites such as the cuticle waxes*

Sample	DDT-RCH (p.p.b.)	Endrin (p.p.b.)
Baled hay residue before benzene washes	395	900
Plant material residues after benzene wash	16	44
Residue removed	96%	95%
Residues total in seven benzene washes [b]	419	998
Total residues recovered in plant material and benzene washes	435	1042
Original residues recovered	110%	108%

[a] Residues calculated on a dry weight basis.
[b] Washed three minutes at 80° C. for each of seven washes.

adding water in excess to cover the plant material, heating to dryness, and then heating additionally for a total of 12 hours.

The effect of oven heat on DDT-RCH and endrin residues on ordinary, field-cured hay is shown in Table VII. When field-cured

Table VII. *Effect on dry and wet oven heat* [a] *on DDT-RCH and endrin residue removal from alfalfa plant material*

Sample	DDT-RCH (p.p.b.)	Endrin (p.p.b.)
Composited baled hay		
No water saturation, no oven heat	430 [b]	900 [b]
No water saturation, dry oven heat	284	587
Loss due to heat	34.0%	35.0%
Composited baled hay		
Water saturated, wet oven heat	58	243
Loss due to heat	86.5%	73.0%

[a] Oven heated 12 hours at 100° C.; residues calculated on a dry weight basis.
[b] Ten percent moisture.

hay of nine to 16 percent moisture was heated in dry oven heat for 12 hours, only 34 percent of the DDT-RCH and 35 percent of the endrin residues were removed. When five replicates of hay containing each pesticide (DDT-RCH and endrin) were saturated with water and were then oven heated for 12 hours, approximately 87 percent of the DDT-RCH and 73 percent of the endrin residues were removed.

The presence of excess moisture in the samples was essential for the effective removal of the pesticide contaminants. After the first treatment with excess water, two similar succeeding treatments did not further reduce the residue levels in the samples.

The effect of dry and wet oven heat on Kelthane residue removal from almond hulls is shown in Table VIII. Dry oven heat at 100°C.

Table VIII. *Effect of dry and wet oven heat [a] on Kelthane residue [b] removal from almond hulls*

Sample	Kelthane (p.p.b.)	4,4'-D[c] (p.p.b.)
Almond hulls		
No water saturation, no oven heat	10,200	N.D. [d]
No water saturation, dry oven heat	3,500	2,700
Loss due to treatment, Kelthane	66%	—
Loss due to treatment, calc. as Kelthane	32%	—
Almond hulls		
Water saturated, wet oven heat	2,300	1,400
Loss due to treatment	77%	—
Loss due to treatment, calc. as Kelthane	59%	—

[a] Oven heated 12 hours at 100° C.
[b] All residues calculated on a dry weight basis.
[c] 4,4'-D signifies 4,4'-dichlorobenzophenone.
[d] N.D. = non-detectable.

for 12 hours removed approximately 66 percent of the original Kelthane residue, but some 4,4'-D which was derived from Kelthane was found on the hulls. If the 4,4'-D was calculated as Kelthane and added to the Kelthane found after treatment, 32 percent of the original Kelthane was removed. The water-saturated hulls after treatment had a 77 percent loss of Kelthane residue but again some 4,4'-D residue was derived from the Kelthane. If the 4,4'-D residue were converted to Kelthane and added to the Kelthane residue found, 59 percent of the original Kelthane residue on the hulls was removed by the oven heating of the almond hulls in the presence of excess moisture.

The effect of dry and wet oven heat on malathion residue removal from ladino seed crop screenings is shown in Table IX. Both dry and wet heatings at 100°C. for 12 hours were effective in removal of the malathion contaminant. Dry heating removed 95 percent and wet heating removed 91 percent of the original malathion, and it appears that both heat treatments were equally effective.

d) Solvent vapor treatment

The dependence of DDT-RCH, endrin, and Kelthane losses on the

Table IX. *Effect of dry and wet oven heat* [a] *on malathion residue* [b] *removal from ladino clover seed crop screenings*

Sample	Malathion (p.p.b.)	Malaoxon (p.p.b.)
Seed crop screenings		
No water saturation, no oven heat	40,000	N.D. [c]
No water saturation, dry oven heat	2,100	N.D. [c]
Loss due to treatment	95%	—
Seed crop screenings		
Water saturated, wet oven heat	3,800	N.D. [c]
Loss due to treatment	91%	—

[a] Oven heated 12 hours at 100° C.
[b] All residues calculated on a dry weight basis.
[c] N.D. = non-detectable.

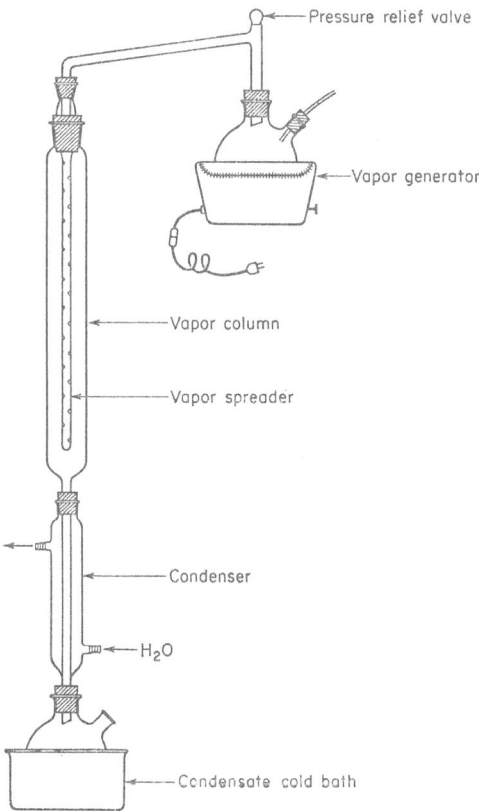

Fig. 1. Apparatus for vapor washing alfalfa hay with aqueous and organic solvents (see text for dimensions) (all joints are $\mathbf{T}$)

Table X. *Removal of DDT-RCH and endrin residues from alfalfa hay* a *by solvent vapor washes*

Sample	Pentane-DDT-RCH (p.p.b)	Benzene-DDT-RCH (p.p.b)	Isopropyl alcohol-DDT-RCH (p.p.b)	Water-DDT-RCH (p.p.b)	Benzene-endrin (p.p.b)	Water-endrin (p.p.b)
Before treatment	560	560	560	560	900	900
After vapor washing	153	16	42	81	64	64
Loss of residue by vapor washing	73%	97%	93%	86%	93%	93%
Recovered in four 500-ml. vapor condensates	426	515	479	31	887	245
Recovered in washed hay	153	16	42	81	64	64
Total recovered	579	531	521	112	951	309
Original residue recovered in washed hay and vapor wash condensates	101%	95%	93%	20%	106%	34%

a Moisture content was approximately 10 percent before treatment; residues calculated on a dry weight basis.

Table XI. Removal of Kelthane residues [a] from almond hulls by solvent vapor washes

Sample	Benzene		Water		15% NH$_3$ in water	
	Kelthane (p.p.b.)	4,4'-D [b] (p.p.b.)	Kelthane (p.p.b.)	4,4'-D (p.p.b.)	Kelthane (p.p.b.)	4,4'-D (p.p.b.)
Before treatment	10,200	N.D.[c]	10,200	N.D.[c]	10,200	N.D.[c]
Vapor washed	900	1,200	2,000	900	4,400	4,600
Loss	91%	—	80%	—	57%	—
Loss calc. as Kelthane	75%	—	68%	—	±	—
Recovered in four vapor condensates	3,900	1,700	N.D.[c]	550	N.D.[c]	190
Total recovered	4,800	2,900	2,000	1,450	4,400	4,790
Total recovered calc. as Kelthane	8,810	—	4,000	—	11,010	—
Recovery of original residue calc. as Kelthane	86%	—	39%	—	108%	—

[a] All residues calculated on a dry weight basis.
[b] 4,4'-D signifies 4,4'-dichlorobenzophenone.
[c] N.D. = non-detectable.

presence of an excess of water vapor suggested that some practical form of "steam distillation" with water or even organic solvents might facilitate removal. The vapor treatment of the plant material was accomplished in a glass apparatus (overall dimensions, 2.1 meters) as shown in Figure 1. The solvent vapors were generated in the three-liter boiler and passed from the spreader through the hay packed in the vapor column (one meter). The entire apparatus, with the exception of the Liebig condenser, was insulated with glass wool and aluminum foil to maintain the solvents in vapor form. The vapors after passing through the plant materials were condensed and collected in 500 ml. fractions in the cold trap. The plant material was washed with the vapors from sufficient solvent to result in four 500-ml. condensate fractions. The water vapor condensates were extracted with benzene before cleanup and analysis.

Table X describes the results of an experiment in which alfalfa hay contaminated with DDT-RCH was treated by vapor washing with pentane, benzene, isopropyl alcohol, and water and, also, hay contaminated with endrin residues was vapor treated with benzene and water. DDT-RCH was removed from the hay in the amounts of 73 percent for pentane, 97 percent for benzene, 93 percent for isopropyl alcohol, and 86 percent for water, while endrin residues were removed in the amounts of 93 percent for benzene and 93 percent for water. The sum of the residues in the vapor-treated plant material and those in the condensates for each organic solvent was essentially equal to the residues on the starting material. The condensates from the water vapor which had passed through the hay were analyzed, but very little pesticide was found. Other workers (BOWMAN et al. 1959) have found that DDT very readily codistills with water, and loss from the receiver despite its low temperature may account for the poor recovery of the pesticides in the water vapor experiments.

The removal of Kelthane residues from the almond hulls by solvent vapor washes is shown in Table XI. Benzene removed 91 percent, water removed 80 percent, and 15 percent ammonia in water removed 57 percent of the Kelthane residues after vapor treatment. In all treated samples, some 4,4'-D residue was derived from the Kelthane present. When the 4,4'-D residues were converted to Kelthane and added to the Kelthane detected after treatment, benzene removed 75 percent, water removed 68 percent, but 15 percent ammonia in water did not show any loss of the original Kelthane on the hulls. If all residues found after treatment including those in the vapor wash condensates were totaled, 86 percent of the original residues were recovered for the benzene vapor treatment, 39 percent for the water vapor, and 108 percent for the 15 percent ammonia in water vapor. Again the low recovery of Kelthane residues from the water vapor treatment was probably due to the volatility of the pesticide in the water vapors.

Table XII. Removal of malathion residues [a] from ladino clover seed crop screenings by solvent vapor washes

Sample	Benzene		Water		15% Ammonia in water	
	Malathion (p.p.b.)	Malaoxon (p.p.b.)	Malathion (p.p.b.)	Malaoxon (p.p.b.)	Malathion (p.p.b.)	Malaoxon (p.p.b.)
Before treatment	40,000	N.D.[b]	40,000	N.D.[b]	40,000	N.D.[b]
Vapor washed	2,900	N.D.	N.D.	N.D.[b]	140	N.D.[b]
Loss	93%	—	100%	—	99%	—
Recovered in four vapor condensates	33,600	N.D.[b]	18,000	N.D.[b]	3,700	N.D.[b]
Total recovered	36,500	—	18,000	—	3,840	—
Recovery of original residue calc. as malathion	91%	—	45%	—	10%	—

[a] All residues calculated on a dry weight basis.
[b] N.D. = non-detectable.

The removal of malathion residues from ladino clover seed crop screenings by solvent vapor washes is shown in Table XII. Benzene removed 93 percent, water removed 100 percent, and 15 percent ammonia in water removed 99 percent of the original malathion residues after vapor treatment. No malaoxon residues were detected in any of the samples. If all the residues found after treatment, including those in the vapor wash condensates and the treated plant material, were totaled, 91 percent of the original malathion residues were recovered for the benzene vapor treatment, 45 percent for the water vapor, and 10 percent for the 15 percent in water vapor.

e) Commercial dehydration

In cooperation with a commercial dehydrator company, a comparison was made between laboratory and commercial samples with respect to DDT-RCH residue loss from alfalfa green-chop of 84 percent moisture content during dehydration. As shown in Table XIII,

Table XIII. *DDT-RCH residue [a] loss from alfalfa green chop during air drying and commercial dehydration*

Sample	Moisture when analyzed (%)	DDT-RCH (p.p.b.)	Loss (%)
Green chop	84	213	—
Air dried [b]	10	111	48
Dehydrated meal [c]	5	96	55
Dehydrated pellets	6	97	55

[a] Residues calculated on a dry weight basis.
[b] Air dried at room temperature in a laboratory hood for five days.
[c] Dehydrated three to five minutes at inlet temperatures of 760° to 1,000° C. and outlet temperatures of 127° C.

green-chop which initially contained 213 p.p.b. of DDT-RCH underwent approximately a 50 percent loss of residue when dehydrated commercially. The residue values, expressed as mean p.p.b. in the air dried, dehydrated meal, and pelleted samples in comparison with the mean p.p.b. values in the green-chop, all showed a significant loss due to dehydration.

f) Hot chemical wash treatment

The almond hull meal and ladino clover seed crop screenings samples were subjected to six separate hot chemical washes for 15 minutes with agitation at a temperature of 60° to 70°C. Separate treatments of the plant samples consisted of 100 percent water, 100 percent ethyl alcohol, two percent potassium hydroxide in 98 percent ethyl alcohol, two percent sodium methoxide in 98 percent ethyl alcohol, two percent sodium ethoxide in 98 percent ethyl alcohol, and two percent sodium ethoxide in 98 percent water. The plant materials were cooled in ice baths and centrifuged from the solutions after the wash treatment. The plant materials were washed twice with 150 ml. of water, and the liquid was separated by centrifugation and pooled with the treatment solutions. After air drying, the plant materials and the liquids were extracted separately with three benzene refluxes, and the samples were stored for analysis.

The results for the almond hull meal are shown in Table XIV. When the treatment solutions consisted of water, 63 percent of the Kelthane residue was removed; ethyl alcohol, 97 percent was removed; sodium ethoxide in water, 81 percent was removed; sodium ethoxide in ethyl alcohol, sodium methoxide in ethyl alcohol, and potassium hydroxide in ethyl alcohol removed 100 percent of the Kelthane. In all cases 4,4'-D residues were detected on the treated plant material. When the 4,4'-D residues were converted to Kelthane and added to the Kelthane residue detected after treatment, total residues removed from the plant material by the treatment were 41, 96, 45, 88, 91, and 87 percent, respectively, as discussed above. Although sodium ethoxide could have been extensively hydrolyzed by the water because the acidity of water and alcohol is about the same, some difference did exist between this sample and the others investigated.

The results for the chemical wash treatment of malathion residues on contaminated ladino clover seed crop screenings are shown in Table XV. No malathion residues could be detected in the treated screenings except in the 100 percent water and 100 percent ethyl alcohol treatments. The water treatment removed 82 percent of the original malathion residue and nine percent was detected in the treatment liquid. The ethyl alcohol treatment removed 89 percent of the original residue and 23 percent was detected in the treatment liquid. All other treatments destroyed 100 percent of the original residue and no malaoxon could be detected in any of the samples.

g) Feed ingredient analyses on untreated and treated samples

To determine whether or not detrimental effects occurred during sample treatments, standard feed ingredient analyses of percent crude protein, crude fat, crude fiber, and ash and, in some instances, vitamin A activity units produced from carotene content/lb. were determined in most of the samples. Table XVI shows the results of standard feed

Table XIV. Chemical treatment of Kelthane-contaminated almond hulls for residue * removal

Sample treatments [b]	Before treatment		After treatment		Removed from hulls as Kelthane (%)	Total removed from hulls calc. as Kelthane (%)	Found in treatment liquids	
	Kelthane (p.p.b.)	4,4'-D [c] (p.p.b.)	Kelthane (p.p.b.)	4,4'-D (p.p.b.)			Kelthane (p.p.b.)	4,4-D (p.p.b.)
Water	10,200	N.D.[d]	3,800	1,600	63	41	170	N.D.
Ethyl alcohol	10,200	N.D.[d]	300	100	97	96	3,700	810
Sodium ethoxide and water	10,200	N.D.[d]	1,900	2,700	81	45	740	220
Sodium ethoxide and ethyl alcohol	10,200	N.D.[d]	N.D.[d]	900	100	88	740	480
Sodium methoxide and ethyl alcohol	10,200	N.D.[d]	N.D.[d]	700	100	91	1,500	1,500
Potassium hydroxide and ethyl alcohol	10,200	N.D.[d]	N.D.[d]	1,000	100	87	N.D.[d]	3,100

a All residues calculated on a dry weight basis.
b Heated 15 minutes at 60° to 70° C. with agitation, cooled, centrifuged, washed with 150 ml. of water. See text for percentage compositions.
c 4,4'-D signifies 4,4'-dichlorobenzophenone.
d N.D. = non-detectable.

Table XV. *Chemical treatment of malathion-contaminated ladino clover seed crop screening for residue [a] removal*

Sample treatments [b]	Before treatment		After treatment		Removed from screenings as malathion (%)	Found in treatment liquids	
	Malathion (p.p.b.)	Malaoxon (p.p.b.)	Malathion (p.p.b.)	Malaoxon (p.p.b.)		Malathion (p.p.b.)	Malaoxon (p.p.b.)
Water	40,000	N.D.[c]	7,200	N.D.[c]	82	3,600	N.D.[c]
Ethyl alcohol	40,000	N.D.[c]	4,500	N.D.[c]	89	9,100	N.D.[c]
Sodium ethoxide and water	40,000	N.D.[c]	N.D.[c]	N.D.[c]	100	N.D.[c]	N.D.[c]
Sodium ethoxide and ethyl alcohol	40,000	N.D.[c]	N.D.[c]	N.D.[c]	100	N.D.[c]	N.D.[c]
Sodium methoxide and ethyl alcohol	40,000	N.D.[c]	N.D.[c]	N.D.[c]	100	N.D.[c]	N.D.[c]
Potassium hydroxide and ethyl alcohol	40,000	N.D.[c]	N.D.[c]	N.D.[c]	100	N.D.[c]	N.D.[c]

[a] All residues calculated on a dry weight basis.
[b] Heated 15 minutes at 60° to 70° C. with agitation, cooled, centrifuged, washed with 150 ml. of water. See text for percentage compositions.
[c] N.D. = non-detectable.

Table XVI. Feed ingredient analysis for alfalfa plant material

Sample	Crude protein (%)	Crude fat (%)	Crude fiber (%)	Ash (%)	Vitamin A activity units from carotene/lb.
Dehyd. alfalfa leaf meal	21.0	3.6	21.0	10.4	150,000
Dehyd. alfalfa meal	17.8	3.0	25.0	9.0	100,000
Sun-cured alfalfa meal	15.2	2.3	28.0	8.5	30,000
Untreated field-cured baled hay	20.4	1.6	27.8	9.3	5,000
Steam vapor-treated baled hay	20.4	1.7	35.3	6.4	5,000
Steam condensates from vapor treatment					
Fraction #1	0.13	N.D.[a]	N.D.[a]	N.D.[a]	N.D.[a]
Fraction #2	0.38	N.D.[a]	N.D.[a]	N.D.[a]	N.D.[a]
Lab. air-dried hay					
Steam-vapor treated	24.4	2.2	29.4	6.3	5,000
No treatment	23.7	2.4	22.1	8.8	13,000
Steam condensates from vapor treatment					
Fraction #1	0.07	N.D.	N.D.	N.D.	N.D.
Fraction #2	0.08	N.D.	N.D.	N.D.	N.D.
Fraction #3	0.27	N.D.	N.D.	N.D.	N.D.

[a] N.D. = non-detectable.

Table XVII. *Feed ingredient analyses for almond hull meal*

Uncoated	Coated	Sample treatment	Crude protein (%)	Crude fat (%)	Crude fiber (%)	Ash (%)	Vitamin A activity units from carotene/lb.
+	−	None	5.4	2.5	13.4	7.3	N.D.[a]
−	+	None	5.5	2.7	14.6	7.1	N.D.[a]
−	+	Refluxed in benzene 1.5 hr.	5.5	1.4	13.2	6.7	N.D.[a]
+	−	Refluxed in benzene 1.5 hr.	5.5	1.4	13.6	8.4	N.D.[a]
−	+	Refluxed 1 hr. in 15% NH_3 in H_2O	10.0[b]	2.2	30.6	5.4	3332
−	+	Refluxed 1 hr. in 20% KOH in EtOH	2.8	0.5	42.7	11.0	1666
−	+	Agitated[c] in 2% KOH and 98% EtOH	5.3	1.5	15.7	10.3	—
−	+	Agitated[c] in 2% NaOEt and 98% EtOH	5.4	1.8	16.3	7.8	—
−	+	Agitated[c] in 2% NaOMe and 98% EtOH	4.7	1.3	15.5	8.8	—
−	+	Agitated[c] in 2% NaOEt and 98% H_2O	5.0	2.0	20.7	10.9	—
−	+	Agitated[c] in 100% H_2O	5.5	1.7	16.2	6.0	—
−	+	Agitated[c] in 100% EtOH	5.3	2.6	17.5	6.4	—
−	+	Benzene-vapor treated	5.7	1.9	13.0	7.5	1666
−	+	Steam-vapor treated	6.4	1.5	23.1	6.8	1666
−	+	Ammonia-vapor treated	17.1[b]	2.3	26.7	3.7	1666
−	+	Three 15 min. washes in H_2O (25° C.)	6.3	3.1	21.2	6.5	—
−	+	Three 15 min. washes in H_2O (90° to 100° C.)	5.8	2.7	23.6	4.7	—
−	+	Dry oven heat at 100° C. for 12 hr.	5.6	2.2	17.0	7.2	—
−	+	Water sat., oven heated at 100° C. for 12 hrs.	5.8	3.3	16.1	7.8	—

[a] N.D. = non-detectable.
[b] High percent protein probably due to residual ammonia.
[c] Stirred for 15 minutes at 60° to 70° C.

Table XVIII. Feed ingredient analyses for ladino clover seed crop screenings

Uncoated	Coated	Sample treatment	Crude protein (%)	Crude fat (%)	Crude fiber (%)	Ash (%)	Vitamin A activity units from carotene/lb.
+	—	No treatment	15.9	2.1	18.9	12.8	5,000
—	+	No treatment	15.1	1.0	21.1	11.5	—
—	+	Benzene-vapor treated	16.3	0.7	20.4	13.6	N.D.[a]
—	+	Steam-vapor treated	13.3	0.7	23.0	11.3	3,333
—	+	Benzene refluxed 1.5 hr.	16.0	0.7	21.3	11.4	N.D.[a]
—	+	Ammonia-vapor treated	15.9	0.9	28.1	12.3	—
—	+	Oven-heated dry	16.0	0.6	21.2	16.5	—
—	+	Oven-heated wet	12.5	1.2	15.3	11.9	—
—	+	Water-washed cold	13.8	0.3	26.8	10.1	—
—	+	Water-washed hot	13.8	0.3	27.6	11.2	—
—	+	Agitated [b] in 2% KOH and 98% EtOH	13.7	0.4	21.3	18.6	—
—	+	Agitated [b] in 2% NaOEt and 98% EtOH	13.4	0.2	20.8	16.1	—
—	+	Agitated [b] in 2% NaOMe and 98% EtOH	13.5	0.2	20.7	18.2	—
—	+	Agitated [b] in 2% NaOEt and 98% H_2O	11.9	0.4	22.3	21.4	—
—	+	Agitated [b] in 100% H_2O	10.5	0.8	11.7	13.9	—
—	+	Agitated [b] in 100% EtOH	13.6	0.8	21.0	14.2	—

[a] N.D. = non-detectable.
[b] Stirred for 15 minutes at 60° to 70° C.

analyses on treated and untreated alfalfa plant material. No significant differences are apparent between non-vapor treated and vapor treated alfalfa hay except in the vitamin A activity units derived from carotene per pound measurements. Steam vapor treatment did reduce the vitamin A activity but had very little effect on the percentages of crude protein, crude fat, crude fiber, and ash.

Table XVII shows the feed ingredient analyses for the almond hull meal. No great differences were observed between the samples except in the refluxed sample in 20 percent potassium hydroxide in ethyl alcohol. The protein value was low and the sample was badly hydrolyzed. The high protein values in the ammonia treated samples were probably due to residual ammonia in the samples. Any reported vitamin A activity in the samples is probably doubtful.

Table XVIII shows the feed ingredient analyses for the ladino clover seed crop screenings. No extreme differences were observed between the samples. However, some of the treatments had some effect on the samples by slightly lowering the percentages of crude protein and crude fat content. Any reported vitamin A activity is probably doubtful.

Acknowledgments

We wish gratefully to acknowledge the help and cooperation of the *California State Department of Agriculture* Feed and Livestock Remedy Laboratory, especially Mr. Van Entwhistle, Mr. James Helmer, and Mr. C. A. Luhman. We also acknowledge the technical assistance of Mr. Stephen Bettcher, Mr. Michael Gilmer, and Mr. Brian Legakis.

Summary

Pesticides have become essential to the economical production of most animal feedstuffs, including also the by-products of other agricultural industries (almond hulls, seed screenings, etc.). The treatment which results in contamination may be intentional (e.g., acaricide application to almond hulls) or incidental (e.g., drift onto alfalfa hay), but an increasing proportion of feed samples contain unacceptable levels of pesticide residues. Removal of these residues may be accomplished by exploitation of some characteristic physical or chemical property such as volatility, instability to hydrolysis, etc. Chlorinated hydrocarbon insecticides have been predominantly responsible for feed contamination; DDT and endrin levels may be substantially reduced in alfalfa products, for instance, by vapor washing or volatilization. Practical-scale examples of the removal of residues of these insecticides and related compounds indicates that decontamination methods demonstrated in the laboratory may become economically feasible.

Résumé *

La décontamination des aliments pour le bétail

Les pesticides sont devenus essentiels pour la production économique de la plupart des aliments pour le bétail, y compris les sous-produits des autres industries agricoles (coques d'amandes, criblage des graines, etc. . .). Le traitement qui provoque une contamination peut être intentionnel (par exemple, le traitement des coques d'amandes par un acaricide), ou accidentel (par exemple, entraînement par le vent sur la luzerne), mais une proportion croissante d'échantillons de fourrages contient des résidus de pesticides à des teneurs inacceptables. L'élimination de ces résidus peut être accomplie en mettant à profit quelques propriétés physiques ou chimiques caractéristiques comme la volatilité, la sensibilité à l'hydrolyse, etc. . . . Les insecticides organochlorés ont une part prépondérante dans la contamination des fourrages. Les teneurs en zeidane et endrine, dans les luzernes, peuvent être substantiellement réduites; par exemple, par lavage à la vapeur ou par volatilisation. Des exemples à une échelle industrielle de la suppression des résidus de ces insecticides et des composés voisins montrent que les méthodes de décontamination prévues au laboratoire peuvent devenir rentables.

Zusammenfassung **

Die Dekontamination von Tierfutter

Pestizide sind für die wirtschaftliche Produktion fast allen Tierfutters, einschliesslich auch der Nebenprodukte von anderen landwirtschaftlichen Industrien (Mandelschalen, Samenauslese, usw.) notwendig geworden. Die Behandlung, die sich dann als Kontamination auswirkt, kann absichtlich sein (z. B. Akarizidbehandlung von Mandelschalen) oder zufällig (z. B. Trift auf Alfalfa Heu), jedoch ein steigender Anteil von Futterproben enthält unannehmbare Mengen von Pestizidrückständen. Entfernung dieser Rückstände kann durch Ausnützung einiger charakteristischer physikalischer oder chemischer Eigenschaften wie Flüchtigkeit, Instabilität bei Hydrolyse usw. erreicht werden. In der Hauptsache sind chlorierte Kohlenwasserstoffinsektizide für die Futterkontamination verantwortlich, DDT und Endrin können wirkungsvoll in Alfalfaprodukten reduziert werden, zum Beispiel durch Dampfwaschung oder Verflüchtigung. Beispiele von praktischem Ausmass für die Entfernung dieser Insektizidrückstände und Rückstände verwandter Verbindungen zeigen, dass die Dekontaminationsmethoden, welche im Laboratorium aufgezeigt werden, wirtschaftlich durchführbar werden könnten.

---

* Traduit par R. Mestres.
** Übersetzt von A. Schumann.

References

ACREE, F., JR., M. BERGER, and M. C. BOWMAN: Codistillation of DDT with water. J. Agr. Food Chem. 11, 278 (1963).

ARCHER, T. E., and D. G. CROSBY: Extraction and location of selective chlorinated hydrocarbon residues in alfalfa hay. Bull. Environ. Contamination Toxicol. 2, 191 (1967).

BOWMAN, M. C., F. ACREE, JR., C. H. SCHMIDT, and M. J. BEROZA: Fate of DDT in larvicide suspensions. J. Econ. Entomol. 52, 1038 (1959).

GRAHAM, S. O.: Indophenol blue as a chromogenic agent for identification of halogenated aromatic compounds. Science 139, 835 (1963).

KOVACS, M. F., JR.: Thin-layer chromatography for organo-thiophosphate pesticide residue determination. J. Assoc. Official Agr. Chemists 47, 1097 (1964).

MITCHELL, L. C.: Separation and identification of chlorinated organic pesticides by paper chromatography. XI. A study of 114 pesticide chemicals: technical grades produced in 1957 and reference standards. J. Assoc. Official Agr. Chemists 41, 781 (1958).

Reduction of parathion residue on celery

By

Neal P. Thompson *

Contents

I. Introduction

We have seen in the past 20 years the rapid growth and expansion in the use of chemicals in the control of insect pests of man and animals and in weed control. Their use is an absolute necessity if we are to continue to attempt to meet the food needs of our time and the daily increasing needs with increasing population. Pesticides, as necessary as we believe they are, provide us with the now well-known concomitant complex problem of residues. The goal toward which we aspire, of course, is that of a pesticide which will perform its work and then disappear entirely in its toxic form from the environment so that no residue problem will exist. One of the more difficult areas of pesticide research is that of removing from a particular commodity any toxic residue, altering the commodity in as small a way as possible. This is not an unexpected difficulty since most pesticide formulations emphasize broad contact and adherence to the sprayed or dusted object for best possible pest control. The particular interest in the research reported here is the attempted removal of parathion from mature celery. Some work was also done with escarole and lettuce which will also be reported.

---

* Department of Food Science, University of Florida, Gainesville, Florida.

II. Previous work

One of the early works in the area of pesticide removal is that of GUNTHER et al. (1950) attempting to remove DDT and parathion residues from several fruits. They stated that none of the experimental treatments afforded significant removal of parathion from treated fruits although they reported some success in removal of DDT. These treatments included such materials as Triton X-100, trisodium phosphate, sodium acid pyrophosphate, and sodium silicate. They suggested that possibly the reasons for lack of significant parathion removal was attributable to subsurface location of the residue. VAN MIDDELEM and WILSON (1955) studied disappearance of parathion from celery under field conditions. Over a period of seven days following the last application, parathion decreased in unwashed celery in the field from 5.03 to 1.05 p.p.m. on foliage and 0.64 to 0.21 p.p.m. on stalks. These residues remained after 15 weekly applications of 1.5 lb. of 15 percent wettable powder/acre. Residues remaining after 12 and nine weekly applications were similar indicating that the significant variable in this case was weathering time in the field after the final spray.

Parathion residues on green beans and mustard greens could be removed 52 and 87 percent, respectively, by neutral soap washing and almost as effectively by Al-ar-polyether alcohol (THOMPSON and VAN MIDDELEM 1955). In these experiments water wash removed 26 and 65 percent, respectively, of the pesticide. They found that with greater initial residues present on the crop greater percentages of parathion could be removed with washing treatment. None of the washing treatments reduced residues to below one p.p.m. GUNTHER et al. (1963) were successful in reducing Guthion residues of 10.5 to 19.0 p.p.m. on oranges.[1] A number of samples were hand washed in dilute Triton X-45, rinsed in distilled water, and air dried. The percent residue reduction by this method ranged from 71 to 96. The residue of Guthion on oranges was reduced from 1.0 to 0.7 p.p.m. by a light chlorine rinse and a germicidal wash containing Ultra wet 60L, trisodium phosphate, and heavy chlorine concentration (ANDERSON et al. 1963). No measurable dimethoate insecticide could be removed by hand washing of Valencia oranges with dilute Triton X-100 (GUNTHER et al. 1965). Apples, plums, and tomatoes were dipped in malathion emulsion or suspension after harvest in preparation for removal studies (KOIVISTOINEN et al. 1964). After varying periods of time following insecticide application (½, 7, and 14 days) the fruits were water washed in a strainer for one minute. Their results for tomatoes are summarized in Table I.

---

[1] Editor's note: GUNTHER has reviewed the removal by washing of 18 insecticides from citrus fruits in the review "Insecticide residues in California citrus fruits and products," Residue Reviews 28, 1 (1969).

Table I. *Removal of malathion from tomatoes by water washing*
(KOIVISTOINEN *et al. 1964*)

Formulation	Age of residue (days)	Malathion (p.p.m.)		Loss (%)
		Before	After	
Suspension	1/2	12.8	2.7	79
	7	5.8	2.8	52
	14	3.5	2.1	40
Emulsion	1/2	3.6	2.3	36
	7	1.4	0.9	36

Notice that final residues after washing changed but a little over the 14-day experimental period when suspensions were used as treatment. These data indicate that a certain amount of malathion applied as suspension became bound even though the prewash residue had decreased markedly.

The chemical and physical nature of plant surfaces in relation to pesticides has been described in considerable detail by CRAFTS and FOY (1962). They reemphasize that the nature of plant surfaces and the composition of the pesticide formulation are both of great importance as factors affecting the penetration of the toxicant. In addition they state that lipoidal constituents of the cuticle constitute an important pool for holding fat-soluble pesticides in solution, restricting further penetration and translocation of such pesticides involving the cuticle itself as a potential residue problem. EBELING (1963) in addition to reviewing the effects of many variables such as the nature of the plant surface, plant growth, formulation, rain, humidity, volatilization, wind, temperature, and light on pesticide residues, refers to subcuticular residues. He places such residues in the category of subsurface residues of GUNTHER *et al.* (1950). EBELING (1963) cites work by LÜDICKE (1949) and JEFFERSON and EADS (1952) which by bioassay indicated that parathion penetrated into leaf tissue in amounts sufficient to kill larvae of various leaf-mining Diptera. EBELING and PENCE (1954) describe experiments showing toxicity to mites placed on the surface of a bean leaf carefully treated only on the opposite surface with parathion. In other experiments (EBELING and PENCE 1953) they reported toxicity to mites one hour after placing them on avocado leaves which had been treated with parathion, washed with detergent solution, and wiped with a rubber sponge. They interpreted these results as indicating that sufficient parathion had penetrated the cuticle to cause toxicity to the mites.

The final report of the *National Canners Association Research Foundation* (LAMB and FARROW 1967) dealing with removal of pesticides from crop material presents data on several fresh crop commodities. Twenty-four percent of parathion residue was removed from spinach leaves containing 1.5 p.p.m. of parathion by commercial detergent wash. Water washing removed less than 10 percent of the parathion. Hand washing in cold water, such as would be done during home preparative procedures, was reported to remove 39 percent of the parathion residue from spinach. Commercial detergent washing of broccoli reduced by 30 percent the level of parathion residue while home preparative methods were ineffective. Tomatoes containing malathion residues subjected to commercial and home preparative washes had 90 and four percent, respectively, removed by the wash. The large difference in wash effectiveness was attributed to different holding conditions and probable penetration of the cuticle by malathion. Green beans containing 1.12 p.p.m. had residues of 0.05 p.p.m. following a five minute cold water wash, a reduction of 96 percent.

We have given some consideration in our laboratory to a carrier solution which would give good contact and coverage but which could be easily removed when desired. We chose the long chain fatty alcohols tetradecanol and heptadecanol hoping that the pesticide, in this case parathion, would have greater affinity for the carrier and would not penetrate plant tissue to as great an extent. Whether our basic premise is true or not is not known at this time; we did find out that we were not able to remove parathion from plant tissue with any greater facility when the insecticide was applied in a fatty alcohol. We have no data indicating whether the effectiveness of the insecticide was altered or not.

PECK (1948) reported on the instability of parathion under alkaline conditions. In seeking a method of removing or degrading parathion, a method which could be carried out in pH of greater than 10 would be desirable. LOSHADKIN and SMIRNOV (1962) list a number of compounds which act as catalysts in the hydrolysis of organophosphorus compounds. Hydrogen peroxide was reported to have a rate constant of hydrolysis which suggested that it would readily catalyze hydrolysis of parathion. Accordingly we have concentrated our efforts on attempts to reduce parathion on mature crops by use of alkaline hydrogen peroxide treatments.

III. Experimental design and results

a) Laboratory applied pesticide removal

Based upon the premise that parathion might be removed from crop plants under alkaline conditions, experiments were designed using hydrogen peroxide in combination with ammonium hydroxide or potassium hydroxide. Initially one ml. solutions of five µg./ml. of parathion in ethanol were mixed with five ml. each of varying concentrations of hydrogen peroxide (1.9 to 7.5 percent) — potassium hydroxide (0.25 to 2.0 percent) to determine whether parathion could be hydrolyzed. Fifteen minutes after combining solutions, the mixture was extracted with benzene. About 10 percent of the original parathion was detectable in the benzene extract by gas chromatography. The higher concentrations of base peroxide were slightly more effective in parathion removal than the lower ones. There was a slight yellowish color in the aqueous phase indicating the probable presence of p-nitrophenol, although this was not checked further. Similar parathion in ethanol solutions were added to diced celery at a concentration of three p.p.m. The celery was washed with 1.0 percent potassium hydroxide — 2.0 percent hydrogen peroxide solution attempting to remove parathion. The wash in these tests and throughout this report implies submersion for 15 to 20 minutes in the wash solution with occasional stirring. In 13 experiments, the average amount of parathion removed was 40 ± 5 percent. The pH of the wash solution varied between 10 and 12. Many of these experiments also included plain water and detergent washes and separate treatments with both reagents. The combination treatment of alkaline peroxide consistently removed more parathion than other treatments. Benzene was used throughout as the extraction solvent in a ratio to crop weight of four ml. to one g. Cleanup was performed with florosil columns after WESSEL (1967) or by co-sweep distillation. Analysis was performed on an F & M 700 gas chromatograph with pulsed electron-capture detector. The instrumental parameters were as follows: three percent SE 30 on 60/80 mesh Gas Chrom Q column at 180° C. with a flow of 120 ml./minute, five percent argon-methane; detector and injection port temperature was 210° C. Results reported herein rely mainly upon relative amounts of parathion present in washed and unwashed samples.

Parathion was also painted on mature crops. Five 10-g. samples of mature celery rib were painted with one ml. of 60 μg./ml. of parathion (six p.p.m.) and were treated and analyzed as shown in Table II. The

Table II. *Removal of laboratory applied parathion (six p.p.m.) from celery by washing*

Carrier	Residue after washing (p.p.m.)		
	KOH-H_2O_2 wash	H_2O_2 wash	KOH wash
Ethanol	0.80	2.35	—
Heptadecanol	1.75	2.85	2.95

results indicate a reduction relative to water wash of 39 and 66 percent when heptadecanol and ethanol, respectively, were carriers.

A similar experiment was performed where five 10-g. samples of celery rib were painted with 0.25 ml. of 120 μg./ml. of parathion (three p.p.m.) and were treated and analyzed as shown in Table III. This

Table III. *Removal of laboratory applied parathion (three p.p.m.) from celery by washing*

Carrier	Residue (p.p.m.)		
	KOH-H₂O₂ wash	No wash	H₂O₂ wash
Ethanol	0.30	2.10	—
Heptadecanol	0.36	1.90	0.43

experiment resulted in a reduction of 77 percent with the hydrogen peroxide wash, and 81 percent with the alkaline peroxide wash when heptadecanol was the carrier and 86 percent when ethanol was the carrier.

Further, three 20-g. samples of celery rib were painted with 0.5 ml. of 120 μg./ml. of parathion (three p.p.m.) and were treated and analyzed as shown in Table IV. Parathion was reduced by 95 percent

Table IV. *Removal of laboratory applied parathion (three p.p.m.) from celery by washing*

Carrier	Residue (p.p.m.)		Reduction (%)
	KOH-H₂O₂ wash	No wash	
Ethanol	0.1	2.0	95
Heptadecanol	0.6	—	—

on these ribs when ethanol was the carrier.

Tetradecanol or heptadecanol as a carrier for parathion applied in the laboratory to celery ribs showed no advantage in removal.

The washing techniques were also applied to two leafy crops, escarole and lettuce. Six 10-g. samples of escarole were submerged for five minutes in 100 ml. of an aqueous solution containing 5.55 mg./1. of parathion dissolved in 10 ml. of 95 percent ethanol. Each sample removed six ml. of aqueous solution. The escarole was dried on polyethylene sheeting for two hours. Three treated samples were washed in 7.5 percent hydrogen peroxide and 2.8 percent ammonium hydroxide solution. Samples were extracted and results are shown in Table V. These data indicate a reduction of 73 percent as a result of alkaline hydrogen peroxide washing.

One ml. of tetradecanol containing 60 μg./ml. of parathion was painted as uniformly as possible on each of four 10-g. lettuce leaves.

Table V. *Removal of laboratory applied parathion from escarole by washing*

Sample	Residue (p.p.m.)		Reduction (%)
	Unwashed	Basic H_2O_2 washed	
A	6.2	1.8	
B	6.2	1.6	73
C	5.6	1.4	
Av.	6.0	1.6	

The lettuce was allowed to stand for several hours after which leaves were extracted and analyzed, as shown in Table VI. This treatment

Table VI. *Removal of laboratory applied parathion (six p.p.m.) in tetradecanol from lettuce by washing*

Sample	Residue (p.p.m.)		Reduction (%)
	Unwashed	Basic H_2O_2 washed	
A	6.3	0.21	
B	6.4	0.70	93
Av.	6.4	0.46	

resulted in an average reduction of 93 percent by washing.

b) Field-weathered pesticide removal

Three 100-g. samples of celery which had been field-treated with parathion at a rate of two lb. of 15 percent wettable powder/acre were washed, extracted, and analyzed, as shown in Table VII. The

Table VII. *Removal of parathion from field-treated celery by washing*

Sample	Residue (p.p.m.)		Reduction (%)
	$KOH-H_2O_2$ wash	No wash	
A	0.02	—	92
	0.03	0.26	89
B	0.04	—	64
	0.07	0.11	36
C	0.07	0.20	—
	0.08	0.18	—
	0.07	0.26	—
Av.	0.07	0.21	65

effectiveness of the basic hydrogen peroxide wash was variable, removing in these experiments 36 to 92 percent of the parathion.

Stalks of field-treated celery which had been harvested two hours after application of three lb./acre of 15 percent parathion wettable powder on February 26, 1968 were washed, extracted and analyzed shown in Table VIII. A reduction of 21 to 49 percent was achieved in this experiment.

Table VIII. *Removal of parathion from field-treated celery by washing*

Elapsed days	Residue (p.p.m.)		Reduction (%)
	KOH-H_2O_2 wash	No wash	
1	0.54	0.92	41
3	0.66	0.84	21
14	0.55	1.08	49

c) *Effect of temperature of wash solution on pesticide removal*

In seeking a practical method for removal of parathion from marketable crops, the commercial hydrocooler seems to be a likely place for a pesticide removal wash. Since the temperature of the wash is below what is normally considered room temperature (72° F.) we compared pesticide removal at 72° F. with that at 35° F.

One ml. of 75 μg./ml. of parathion in ethanol was added to 25-g. portions of celery rib and the ribs were set aside to dry for about 16 hours. Celery ribs and wash solutions were kept at the prescribed temperatures during the course of the experiment. The ribs were washed, extracted, and analyzed. Reductions of parathion on celery at room temperature and at 35° F. are shown in Table IX.

Table IX. *Removal of laboratory applied parathion (three p.p.m.) from celery at wash temperatures of 72° and 35° F.*

Treatment	Residue (p.p.m.) at		Reduction (%)	
	72° F.	35° F.		
No wash	1.96	1.58	—	—
H_2O wash	1.72	1.48	12	6
KOH [a] wash	1.66	1.52	15	4
H_2O_2[b] wash	1.70	1.52	13	4
H_2O_2[c]-KOH [d] wash	1.18	1.10	40	30

[a] Four percent.
[b] Seven percent.
[c] 3.5 percent.
[d] Two percent.

Field-treated celery was also washed at 72° and 35° F. with 3.5 percent hydrogen peroxide — 2.0 percent potassium hydroxide. To reduce variability, stalks were quartered and two quarters were washed and two were not. Results are shown in Table X. Comparison

Table X. *Removal of parathion from field-treated celery at wash temperatures of 72° and 35° F.*

Temperature (° F.)	Residue (p.p.m.)		Reduction (%)
	No wash	H$_2$O$_2$-KOH wash	
72	1.04	0.49	57
	1.10	0.44	
	0.70	0.12	53
	0.81	0.59	
35	0.86	0.50	40
	0.85	0.53	

of 72° and 35° F. wash treatments shows that the lower temperature wash is effective but a smaller percentage removal of parathion is achieved.

One of the major concerns in research of this type is the phytotoxicity of the wash solution. We have not looked into this aspect intensively as far as storage or shelf life is concerned. We have noticed a water-soaked appearance of the celery which has been treated with the higher concentrations of alkaline wash. This appearance evidently is similar to that reported by other workers for detergent washes. The lower concentrations of alkaline wash used in the experiments forming a major portion of those reported here indicated no differences in phytotoxicity relative to plain water-washed samples. Taste tests performed according to the triangle flavor difference test showed any apparent difference in taste as a result of the wash to be insignificant.

The results reported, while by no means exhaustive, demonstrate a method for use in reduction of parathion residue on celery and also on lettuce and escarole.

Summary

The removal of pesticides from mature crops is a challenging task because of formulation design in pesticide chemicals and the nature of plant surfaces. A method is described which employs an alkaline hydrogen peroxide wash treatment of parathion contaminated mature celery, escarole and lettuce. The percentage of parathion removed by this wash is variable ranging from 20 to above 90 percent. The variability is dependent somewhat on whether parathion had been applied in the laboratory or in the field and on the concentration of the wash ingredients. Parathion could be removed at wash temperatures of 72° and 35° F. with the lower temperature being slightly less effective.

Phytotoxicity was limited to high concentrations of wash and taste tests showed no significant differences between wash treated and untreated check plant material.

Résumé *

Réduction des résidus de parathion sur céleris

L'élimination des pesticides des récoltes à maturité constitue un défi, en raison de la composition des formulations des pesticides chimiques et de la nature des surfaces végétales. On décrit une méthode basée sur le lavage du parathion des céleris, des scaroles et des laitues pleinement développées, à l'aide de peroxyde d'hydrogène alcalin. Le pourcentage de parathion éliminé par ce lavage est variable ; il se range entre 20 et plus de 90 pour cent. Il varie quelque peu selon que le parathion a été appliqué en laboratoire ou en plein champ et selon la concentration des produits de la solution de lavage. Le parathion peut être éliminé aux températures de lavage de 72° et 35° F, la température la plus basse étant légèrement moins efficace. La phytotoxicité s'est limitée aux concentrations élevées des solutions de lavage ; les tests organoleptiques n'ont montré aucune différence significative entre les produits végétaux traités par lavage et les témoins.

Zusammenfassung **

Reduzierung von Parathionrückständen auf Sellerie

Die Entfernung von Pestiziden von reifen Feldfrüchten ist eine herausfordernde Aufgabe wegen der Art der Formulierung von Pestizidchemikalien und der Natur der Pflanzenoberflächen. Es wird eine Methode beschrieben, welche eine alkalische Wasserstoffperoxyd-waschbehandlung an Parathion behandeltem reifen Sellerie, "Escarole" und Salat anwendet. Der Prozentsatz an Parathion, der mit dieser Waschung entfernt wird, ist unterschiedlich und rangiert von 20 bis über 90 Prozent. Die Streuung ist etwas abhängig davon, ob Parathion im Labor oder im Feldversuch appliziert wurde und andererseits von der Konzentration der Bestandteile der Waschlösung. Parathion konnte bei einer Waschtemperatur von 72° und 35° F entfernt werden, wobei die niedrigere Temperatur etwas weniger wirkungsvoll war. Pflanzentoxizität war beschränkt auf hohe Konzentrationen von Waschlösung, und Geschmackstestе zeigten keine bemerkenswerten Unterschiede zwischen waschbehandeltem und unbehandeltem Kontrollpflanzenmaterial.

---

* Traduit par S. DORMAL-VAN DEN BRUEL.
** Übersetzt von A. SCHUMANN.

References

ANDERSON, C. A., D. MACDOUGAL, J. W. KESTERSON, R. HENDRICKSON, and R. F. BROOKS: The effect of processing on Guthion residues in oranges and orange products. J. Agr. Food Chem. 11, 422 (1963).

CRAFTS, A. S., and C. L. FOY: The chemical and physical nature of plant surfaces in relation to the use of pesticides and to their residues. Residue Reviews 1, 112 (1962).

EBELING, WALTER: Analysis of the basic processes involved in the deposition, degradation, persistence, and effectiveness of pesticides. Residue Reviews 3, 35 (1963).

——, and R. J. PENCE: Pesticide formulation. Influence of formulation on effectiveness. J. Agr. Food Chem. 1, 386 (1953).

—— —— Susceptibility to acaricides of two-spotted spider mites in the egg, larval, and adult stages. J. Econ. Entomol. 47, 789 (1954).

GUNTHER, F. A., M. M. BARNES, and G. E. CARMAN: Removal of DDT and parathion residues from apples, pears, lemons, and oranges. Adv. Chem. Series 1, 137 (1950).

——, G. E. CARMAN, R. C. BLINN, and J. H. BARKLEY: Persistence of residues of Guthion on and in mature lemons and oranges and in laboratory processed citrus "pulp" cattle feed. J. Agr. Food Chem. 11, 424 (1963).

——, W. H. EWART, J. H. BARKLEY, and R. T. MURPHY: Persistence of residues of dimethoate on and in mature Valencia oranges and in laboratory-processed citrus pulp cattle feed. J. Agr. Food Chem. 13, 549 (1965).

KOIVISTOINEN, P., A. KARINPAA, M. KONONEN, and P. ROINE: Malathion residues on fruit treated by dipping. J. Agr. Food Chem. 12, 551 (1964).

LAMB, F. C., and R. P. FARROW: Investigations on the effect of preparation and cooking on the pesticide residue content of selected vegetables. National Canners Association Research Foundation, Final Report, Washington, D.C. (1967).

LOSHADKIN, N. A., and V. V. SMIRNOV: A review of modern literature on the chemistry and toxicology of organophosphorus inhibitors of cholinesterases. Associated Technical Services, Inc., Glen Ridge, New Jersey (1962).

PECK, D. R.: Alkali stability of the insecticide E. 605. Chem. & Ind. P. 526 (1948).

THOMPSON, B. D., and C. H. VAN MIDDELEM: The removal of toxaphene and parathion residues from tomatoes, green beans, celery and mustard with detergent washings. Proc. Amer. Soc. Hort. Sci. 65, 357 (1955).

VAN MIDDELEM, C. H., and J. W. WILSON: Parathion residues on celery. J. Econ. Entomol. 48, 88 (1955).

WESSEL, J. R.: Collaborative study of a method for multiple organophosphorus pesticide residues in nonfatty foods. J. Assoc. Official Anal. Chemists 50, 430 (1967).

Accelerated removal of pesticides from domestic animals

By

B. J. Liska* and W. J. Stadelman*

Contents

I. Introduction

Domestic animals may be exposed to pesticides in a variety of ways. They may consume feed with residues present. These residues in or on feed are absorbed in the intestinal tract. Some of the residues are de-toxified and excreted in feces and urine. In the case of the chlorinated insecticides a portion of the residues are deposited in lipid tissues in the body. A second means of contamination is when animals are treated directly for control of pests with insecticides. If animals are present when spraying for pests is done in buildings or out in the open they may become contaminated by inhaling the spray material or having sprays come in contact with the skin. Another possible method of contamination is for animals to come in contact with surfaces which have been sprayed with chlorinated pesticides and absorb these residues through the skin.

Once the animals or birds are contaminated with insecticides of the chlorinated type they will produce milk or eggs containing residues for extended periods of time. Also meat from the animals or birds will be contaminated following slaughter. Since the tolerance levels for chlorinated pesticide residues in animals and poultry products are set as very low levels this means the opportunity for having products above tolerance levels are numerous.

---

* Food Sciences Institute, Purdue University, Lafayette, Indiana.

The animal and poultry industries have become interested in methods to reduce the levels of chlorinated pesticide residues in animal and poultry products. This can normally be done by monitoring feeds and controlling the environment to keep contamination of animals and birds to a minimum. What of the case where accidental contamination of a herd of dairy cows or a flock of laying hens occurs? Should these animals or birds be destroyed? Products from contaminated animals can be kept off the market until the contamination decreases or dissipates entirely but this requires weeks or months in some cases.

In the past several years activity has increased in research studies on decontamination of animals and birds. This can be approached in several ways. First, by changes in diet to accelerate removal of fat from animals or birds, with accompanying removal of residues. Second, use of chemicals to increase de-toxification of residues ingested. Third, by prevention of residue absorption from the intestinal tract. This paper will review reported work in this general area of intentional decontamination research on domestic animals and birds.

II. Residues in animals and birds

MARTH (1965) summarized information on residues of chlorinated pesticides in animals and poultry. This is an excellent review of sources of contamination and levels of residues resulting from specific amounts of contamination. A report by LISKA et al. (1964 a) indicated possible contamination of turkeys from lindane-sprayed range. LISKA et al. (1964 b) also reported the contamination of broilers and laying hens from feeds contaminated with low levels of chlorinated insecticides. There have been several recent reports of commercial flocks of laying hens contaminated from feeds consumed.

WITT et al. (1966 a) presented information on contamination of milk from cows exposed to DDT in by several different methods. BROWN et al. (1966) and STULL et al. (1966) presented further evidence on presence of residues in milk from contaminated animals. WITT et al. (1966 b) explored the effect of respiratory exposure of dairy animals to pesticides. Results indicated this to be a minor route for entrance of pesticides into milk. STULL et al. (1968 a) reported on an epidemoiological study of DDT contamination in milk on a dairy farm. They reported feed levels in excess of 0.10 p.p.m. gave rise to milk fat levels of DDT over 1.2 p.p.m. LABEN (1968) indicated similar results in reviewing residue problems in dairy production.

One of the first approaches to decontamination of animals or birds is in some way to mobilize the depot fat. In this way some of the pesticide was eliminated and the remaining residues would be further diluted as depot fat was replaced. By reducing the residues to a sufficiently low level so milk or eggs are below tolerance levels the resulting foods could therefore be marketed safely.

III. Decontamination of poultry

STADELMAN (1965) and WESLEY et al. (1966 and 1968) reported on research performed using laying hens. Laying hens were contaminated with DDT by force feeding hens gelatin capsules containing DDT to give a 10 to 15 p.p.m. exposure based on 0.3 lb./day feed consumption. Table I contains data (STADELMAN et al. 1965) on removal

Table I. *Diminution of pesticide residues in depot fat of laying hens during time after exposure to the equivalent of 10 to 15 p.p.m. of the insecticides in the feed for five days* (STADELMAN et al. 1965)

Weeks after exposure	Residue in depot fat (p.p.m.)			
	Lindane	Dieldrin	Heptachlor + epoxide	DDT + DDE
1	0.7	3.6	10.2	9.6
3	0.3	4.9	0.6	6.4
10	0.1	3.9	1.1	1.6
17	0.0	1.0	0.5	1.6
26	0.0	1.0	0.3	0.7

of DDT, lindane, dieldrin, and heptachlor from contaminated hens under control conditions with no special techniques to speed removal. Table II contains data on the decrease in residues in egg yolk in eggs

Table II. *Diminution of pesticide residues in egg yolk during time after exposure to the equivalent of 10 to 15 p.p.m. of the insecticides in the feed for five days* (STADELMAN et al., 1965).

Weeks after exposure	Residue in egg yolk (p.p.m.)			
	Lindane	Dieldrin	Heptachlor + epoxide	DDT + DDE
1	0.5	0.7	1.1	5.1
5	0.1	1.2	1.4	1.7
10	0.1	1.6	0.4	0.5
17	0.0	0.3	0.1	0.5
26	0.0	0.3	0.2	0.2

from the hens over time. There was a difference in persistence of the various insecticides used. After 26 weeks only lindane has decreased to levels below detection limits of the method used. Other insecticide residues used were still present in depot fat and egg yolks (Stadelman *et al.* 1965).

As a part of this experiment Wesley *et al.* (1966) placed groups of the contaminated laying hens on a modified forced molt procedure (*Washington State Poultry Council* 1947). The hens were allowed no feed for 48 hours. Water was available at all times. Following this period the hens were placed either on a low or high protein ration. One group was not starved and used as a control. Tables III and IV

Table III. *DDT and DDE residues in abdominal fat samples of laying hens* (Wesley *et al.* 1966)

Week of sampling	Control	High protein (p.p.m.)	Low protein (p.p.m.)
1	9.4	9.6	9.8
2	8.5	8.8	8.7
3	8.2	8.2	8.3
4	7.0	6.7	6.9
6	5.4	5.2	5.2
8	3.3	2.8	3.1
10	2.5	1.2	1.6
12	1.7	0.4	0.8
17	0.4	0.1	0.2
22	0.1	0.0	0.1

Table IV. *Residues of DDT and DDE in egg yolk as influenced by management of layers* (Wesley *et al.* 1966)

Week of sampling	Control	High protein (p.p.m.)	Low protein (p.p.m.)
1	3.5	3.6	3.5
2	2.9	— [a]	— [a]
4	2.5	— [a]	— [a]
6	2.1	— [a]	2.1
8	1.5	0.5	0.9
10	1.1	0.3	0.7
12	0.9	0.3	0.5
14	0.5	0.1	0.2
16	0.2	0.0	0.1
18	0.1	0.0	0.1
20	0.1	0.0	0.0

[a] The hens were out of production due to management practices.

contain data from Wesley *et al.* (1966) which demonstrated some increase in rate of removal by these techniques. Eggs with less than

0.1 p.p.m. DDT + DDE were available from the contaminated birds on the high protein ration 14 weeks after decontamination started. Those hens on a low protein diet reached this level in 18 weeks while controls required 20 weeks. Abdominal fat samples from the hens still contained detectable residues for several weeks.

In a second series of experiments WESLEY et al. (1968) varied protein content of the ration, used androgen injection, and varied times of starvation at 48 hours and 96 hours (in two 48-hour periods). Results indicated a non-linear depletion rate for DDT over time. High protein rations increased depletion rates of DDT. A single injection of androgen at the time of DDT exposure had no effect on depletion rate. As a part of this series of experiment several groups of laying hens were placed on rations with fat plus vitamins, with fat minus vitamins, low fat with vitamins, and low fat minus vitamins. Table V

Table V. *DDT residues in abdominal fat of laying hens as affected by fat and fat-soluble vitamins in the ration* (WESLEY et al. 1968)

Group [a]	Total residues [b]		Ration fed [d]
	Before DDT exposure	28 Days after first DDT exposure [c]	
1	1.07	18.36	HF – V
2	1.45	18.55	HF + V
3	1.62	13.31	LF – V
4	1.54	11.38	LF + V

[a] The remaining 10 chickens of each group, after five chickens from each group were sacrificed to established residual level of DDT before exposure.
[b] DDT plus its metabolites, given as the mean of each group.
[c] Exposure was to 2.35 mg. of DDT in capsules for five days.
[d] HF – V = 5% added fat, no fat-soluble vitamins added.
  HF + V = 5% added fat, fat-soluble vitamins added.
  LF – V = low-fat (less than 0.01% by manufacturer's analysis), no fat-soluble vitamins added.
  LF + V = fat-free, fat soluble vitamins added.

includes data on results obtained. Hens on the high fat ration retained about 40 percent more DDT residues than those on the low or fat free diet. Those hens on the low fat diet without fat soluble vitamins retained about 15 percent more DDT than those with added vitamins. LISKA et al. (1964 b) fed rations to broilers containing either corn oil or hydrogenated coconut oil as fat sources. The rations were contaminated with 0.1 or 1.0 p.p.m. of DDT. The broilers on the corn oil ration had 20 to 25 percent less DDT residue deposited in body fat (Table VI) than those on the hydrogenated coconut oil ration. This was

Table VI. *DDT residues in abdominal fat and other broiler tissues* [a, b]
(LISKA *et al.* 1964 b)

Sample	Control ration	CO [c] + 0.1 p.p.m. DDT ration	HCO [d] + 0.1 p.p.m. DDT ration	CO [c] + 1.0 p.p.m. DDT ration	HCO [d] + 1.0 p.p.m. DDT ration
Fat	0.2	0.8	1.0	2.30	2.8
Skin	0.1	0.7	0.90	1.3	1.3
Thigh meat	0.1	0.7	0.80	1.1	1.4
Breast meat	0.1	0.5	0.70	0.9	0.8

[a] Average results for four birds/group.
[b] p.p.m. of DDE and DDT combined to nearest 0.1 p.p.m.
[c] CO indicates 10 percent corn oil in ration.
[d] HCO indicates 10 percent hydrogenated coconut oil in ration.

true at both levels of DDT used. These results may indirectly indicate some factors involved in DDT transfer from the intestinal tract.

The results of this work have been applied commercially with some success. There is a low mortality of birds using the forced molt procedure. The hens do go out of production for a few weeks but their production rate is higher when production resumes so total production is quite similar. This could serve as a routine method for reduction of residues in commercial laying flocks if needed in the future.

IV. Decontamination of dairy cows

LABEN (1968) has reviewed the problem of contamination in dairy cattle. The report includes results by LABEN *et al.* (1966) and WITT *et al.* (1966 a) on the dangers of exposure of dairy cows to one lot of contaminated feed. LABEN (1966) also indicates the effects of continuous low level feed contamination on residues in milk.

Some attempts using thyroprotein to speed removal of pesticides from dairy cows have been reported. STULL *et al.* (1968 b) reported that feeding thyroprotein increased milk output and loss of body weight but did not speed up DDT removal from depot fat. In contrast to these results, MILLER (1967) and STREET (1968) indicate that other research has shown a beneficial effect from thyroprotein.

WILSON *et al.* (1968) have reported that rate of pesticide removal from dairy cows was doubled by feeding small quantities of activated charcoal following contamination. If charcoal was fed alone with the contaminated feed much of the DDT passed out in urine and feces instead of being absorbed by the cow. The addition of small amounts of charcoal to feed could serve as a built-in filter mechanism to reduce residues in animals and animal products. There are several questions to answer: What nutrients are removed? What effect would it have on appetite and possibility any other long term effects? Overall the results are encouraging and offering a practical method for control of and removal of residues in ruminants. This principle might be extended to include natural or synthetic adsorbent materials available now or developed in the future.

V. Decontamination research on rats

Another approach for increasing the rate of removal of persistent pesticides from animals is to speed up detoxification or metabolic degradation of the chemicals in vital organs. Enzyme systems in the liver can be induced by DDT to more rapidly detoxify dieldrin. STREET (1968) has covered the reports of research groups that have obtained an increase in the detoxification of chlorinated pesticides in rats after feeding other pesticides or drugs of various types. The barbiturates have served to induce increased enzyme function and more rapid residue removal. COOK and co-workers at *Michigan State University* are actively engaged in this area of research. As pointed out by STREET (1968) this treatment is effective in rats but may not be the case for mammals in general. The real problem centers on finding a chemical that will trigger the induction of enzyme systems but not be a residue or have any side effects. This same procedure might be useful in humans. The drug consumption at present may be helping keep pesticide residues in humans at the present levels.

VI. Decontamination of feeds

STREET (1968) summarized work on decontamination of feeds. He included reports on effects of high temperature dehydration of alfalfa. Significant reduction in chlordane, DDT, toxaphene, heptachlor, parathion, aldrin, and malathion was reported by a number of research groups. Recent studies have shown that low levels of chlorinated hydrocarbon residues in alfalfa can be removed by extraction with organic solvent vapors; over 70 percent removal was achieved in this work by ARCHER and CROSBY (1967).

The question must be asked is this feasible, will flavors and odors from solvent residues be a problem, and how will it affect acceptance of the feed. This is one more approach to help in combination with decontamination procedures previously discussed.

In discussing contamination of animals or poultry and methods for intentional decontamination, it becomes obvious that there is a need for more knowledge about transport of chlorinated pesticides in the biological system. WILSON et al. 1968, WITT et al. (1966 c), and VESSEY et al. (1968) have reported on phases of the problem. It is imperative that information on absorption and transport of chlorinated pesticides in the animal or bird systems be obtained in order to take proper steps to prevent absorption or increase the rate of removal to the maximum. This is information that could be important for "routine" reduction or present levels of persistent pesticides in domestic animals from unavoidable contamination. Further, the procedures will be of extreme importance in cases of accidental contamination of domestic animals in order to reduce economic losses to producers and safeguard our food supply.

Summary

Principal methods of contamination of domestic animals are from contaminated feeds, spraying operations and contact in the environment. Procedures for intentional removal of chlorinated pesticides from poultry and dairy cows are reported. Use of enzyme inducing drugs to de-toxify pesticides is progressing in small animals. This may have application in food producing animals. There is a need for more basic information on pesticide adsorption and transport in biological organisms.

Résumé *

Elimination accélérée de pesticides des animaux domestiques

Les principales sources de contamination des animaux domestiques sont les aliments du bétail contaminés, les opérations de pulvérisation et le contact avec le milieu ambiant. On fait rapport sur des procédés destinés à éliminer intentionnellement les pesticides chlorés de la volaille et des vaches laitières. L'usage de médicaments inducteurs de réactions enzymatiques de détoxification des pesticides est en progrès chez les petits animaux. Ceci peut avoir une application chez les animaux producteurs de denrées alimentaires. Il est nécessaire d'avoir plus d'informations fondamentales sur l'adsorption des pesticides et leur transport dans les systèmes biologiques.

Zusammenfassung **

Beschleunigte Entfernung von Pestiziden in Haustieren

Die Hauptmethoden der Kontamination von Haustieren sind durch kontaminiertes Futter, Spritzbehandlungen und Kontakt mit der Umgebung. Methoden für die vorsätzliche Entfernung von chlorhaltigen Pestiziden in Geflügel und Milchkühen werden beschrieben. Der Ge-

---

* Traduit par S. Dormal-van den Bruel.
** Übersetzt von A. Schumann.

brauch von Enzym-induzierenden Drogen, um Pestizide zu detoxifi-zieren, macht Fortschritte in kleineren Tieren. Diese Methode könnte Anwendung finden in Nahrung produzierenden Tieren. Es besteht die Notwendigkeit von mehr Grundlageninformation von Pestizidabsorp-tion und Transport in biologischen Organismen.

References

ARCHER, T. E., and D. G. CROSBY: Annual progress report, Regional Research Project W-45, CSRA, *U.S. Department of Agriculture* (1967).

BROWN, W. H., J. M. WITT, F. M. WHITING, and J. W. STULL: Secretion of DDT in milk by fresh cows. Bull. Environ. Contamination and Toxicol. 1, 21 (1966).

LABEN, R. C.: DDT contamination of feed and residues in milk. J. Animal Sci. (In press) (1968).

——, T. E. ARCHER, D. G. CROSBY, and S. A. PEOPLES: Milk contamination from low levels of DDT in dairy rations. J. Dairy Sci. 49, 1488 (1966).

LISKA, B. J., G. C. MOSTERT, B. E. LANGLOIS, and W. J. STADELMAN: Problems resulting from the misuse of lindane for chigger control on turkey ranges as related to residue in edible tissues. J. Econ. Entomol. 57, 682 (1964 a).

——, B. E. LANGLOIS, G. C. MOSTERT, and W. J. STADELMAN: Residues in eggs and tissues of chickens on rations containing low levels of DDT. Poultry Sci. 43, 982 (1964 b).

MARTH, E. H.: Residues and some effects of chlorinated hydrocarbon insecticides in biological material. Residue Reviews 9, 1 (1965).

MILLER, D. D.: Effects of thyroprotein and a low energy ration on removal of DDT from lactating dairy cows. J. Dairy Sci. 50, 1444 (1967).

STADELMAN, W. J.: Management practices for removal of residues from domestic animals. Proc. 25th Semi-annual Meeting Amer. Feed Manufacturers Assoc. Nutr. Council, p. 35 (1965).

——, B. J. LISKA, B. E. LANGLOIS, G. C. MOSTERT, and A. R. STEMP: Persistence of chlorinated hydrocarbon insecticide residues in chicken tissues and eggs. Poultry Sci. 44, 435 (1965).

STREET, J. C.: Methods of removal of pesticide residues. J. Can. Med. Assoc. (In press) (1968).

STULL, J. W., W. H. BROWN, F. M. WHITING, and J. M. WITT: Variability of DDT in milk. J. Dairy Sci. 49, 945 (1966).

—— —— —— —— An epidemiological study of pesticide contamination in milk on an operating dairy farm. Bull. Environ. Contamination and Toxicol. (In press) (1968 a).

—— —— ——, L. M. SULLIVAN, MARY MILBRATH, and J. M. WITT: Secretion of DDT by lactating cows fed thyroprotein. J. Dairy Sci. 51, 56 (1968 b).

VESSEY, D. A., L. S. MAYNARD, W. H. BROWN, and J. W. STULL: The subcellular partition and metabolism of orally administered 1,1,1-trichloro-2,2-bis(p-chlorophenyl) ethane by rat liver cells. Biochem. Pharmacol. 17, 171 (1968).

Washington State Poultry Council: Summer forced molting of hens for commercial egg production. Poultry Pointer No. 21, Agr. Extension Bull. 323 (1947).

WESLEY, R. L., A. R. STEMP, B. J. LISKA, and W. J. STADELMAN: Depletion of DDT from commercial layers. Poultry Sci. 45, 321 (1966).

—— ——, R. B. HARRINGTON, B. J. LISKA, R. L. ADAMS, and W. J. STADELMAN: Further studies on depletion of DDT residues from laying hens. Poultry Sci. 47. In press (1969).

WILSON, K. A., R. M. COOK, and R. S. EMERY: Effects of charcoal feeding on dieldrin excretion in ruminants. Fed. Proc. 27, abstr. #1924 (1968).

WITT, J. M., F. M. WHITING, W. H. BROWN, and J. W. STULL: Contamination of milk from different routes of animal exposure to DDT. J. Dairy Sci. 49, 370 (1966 a).

—— —— —— Respiratory exposure of dairy animals to pesticides. Adv. Chem. Series 60, 99 (1966 b).

——, W. H. BROWN, G. I. SHAW, L. S. MAYNARD, L. M. SULLIVAN, F. M. WHITING, and J. W. STULL: Rate of transfer of DDT from the blood compartment. Bull. Environ. Contamination Toxicol. 1, 187 (1966 c).

Effects of processing on pesticides in foods

By

B. J. LISKA* and W. J. STADELMAN*

Contents

I. Introduction

Chlorinated insecticides, since their introduction about 25 years ago, have been used so extensively that now they appear to be a permanent part of man's environment. The length of time they will remain in the environment and find their way into biological systems is not known. As a result of this persistence they are translocated in water, soil, plants, and animals throughout the food chain until they reach man's food supply. MARTH (1965) summarized the information on pesticide residues in biological materials. From his paper one concludes that pesticide residues can be found wherever plant or animal life exists.

Using present food production practices and those envisioned in the near future, we are forced to accept the fact that small quantities of pesticide residues will be present in our food supply. The significance of minute quantities of these chemical residues in the food supply, over an extended period of time, on human health is a much debated question. Can these residues by themselves effect genetic development in future generations? Will there be interactions between these chemicals and other chemicals as food additives, medicines, and other chemicals in man's environment to cause long term genetic

---

* Food Sciences Institute, Purdue University, Lafayette, Indiana.

changes? This becomes one more problem for man to face in controlling the total insult to his environment.

Man's food can serve as a continual source of chemical residues which accumulate in the human body (DURHAM 1963). The *U.S. Department of Agriculture* and *U.S. Food and Drug Administration* have monitored the level of pesticide residues in raw and processed foods for several years. A recent summary of this work (DUGGAN 1968) indicates the base of information available; a part of this study since 1963 was reported by DUGGAN and DAWSON (1967). The data indicated that the average U.S. diet contained the following residues:

Chlorinated organic chemicals	0.02 p.p.m.
Organic phosphate chemicals	0.003 p.p.m.
Carbamate chemicals	0.05 p.p.m.
Chlorophenoxy chemicals	0.003 p.p.m.

These authors noted an increase from 0.08 mg./day to 0.12 mg./day in chlorinated pesticides from 1963 to 1964. Since 1964 the levels of pesticides remained constant. The report by DUGGAN (1968) gave interesting data on which food groups contributed the chlorinated hydrocarbon residues in the total diet. In 1966-67 dairy products contributed 17.5, meat, fish and poultry 35.0, grains and cereals 7.5, potatoes 1.3, leafy vegetables 2.5, garden fruits 10.0, fruits 22.5 and oils, fats, and shortening 2.5 percent of the chlorinated hydrocarbon residues. This indicates which food groups need continued monitoring and further research on production and processing practices in order to maintain present levels or reduce the residue levels in the future.

Many tolerances on raw agricultural and processed food products are now in use. Recently the dairy industry obtained a long sought-after DDT tolerance (*Federal Register* 1967). Indications are that some tolerances are being revised downward. With this as background it becomes obvious that the effects of processing chemical residues in foods is an area where available information should be consolidated and missing information obtained by further research. The information would be useful in establishing tolerances on raw and processed food products and in determining whether foods contaminated above tolerance levels could be processed to reduce the pesticide levels to acceptable limits.

CROSBY (1965) summarized the information on effects of normal processing on residues in foods up to 1964. Some information was available on fruits and vegetables but virtually none on animal products. Reviews by LISKA (1968) and STREET (1968) cover more recent research information. The purpose of this paper is to review the research on effects of normal and special processing techniques on pesticide residues as a means of decontaminating or reducing residue levels in foods when consumed.

II. Location of residues in foods

The type of pesticide involved and the type of food are main factors determining the location of the residue in the food. The chlorinated hydrocarbon insecticides in most cases in animal products are associated with lipid materials (DURHAM 1963). LANGLOIS et al. (1964 and 1965) and LISKA et al. (1967) reported on location of residues in various chicken tissues and in milk fractions. The occurrence of lindane, DDT, dieldrin, heptachlor, and endrin was closely related to lipid content of the tissue or fraction. In some cases this is an advantage in removal of the residue if concentration in one tissue or food component occurs. Milk can be separated into a skim-milk and cream portion with virtually all residue of chlorinated hydrocarbons in the cream portion. Egg yolks from contaminated eggs will contain all the chlorinated insecticide residues with the whites essentially residue free. Organophosphate residues would be located in both fractions of eggs or milk. In some preliminary experiments, milk containing DDT was separated and the cream was further processed into butteroil using a combination detergent and heat processing procedure (STINE and PATTON 1952). From 27 to 53 percent of the DDT was removed during the processing into butteroil in four trials. If the DDT were located on the fat globule membrane one would expect a higher percentage removal. Apparently the DDT is located throughout the fat globule.

In plant materials the residue may be dried on the surface of the raw product, absorbed or bound to waxy materials in skin of fruits or vegetables, or translocated to the inner tissues. As indicated by LICHTENSTEIN et al. (1965) varietal differences may enter into absorption of pesticides from soil by vegetables. In the study involving several carrot varieties inner tissues varied from 14 to 50 percent in amount of adsorbed residue. Also, 30 minute boiling removed heptachlor but not aldrin from the inner tissue into the cooking liquid. BOURNE (1967) reported that adsorption of DDT residues on apple varieties was related to the wax content of the apple skin. ANDERSON et al. (1963) found that washing removed 30 percent of Guthion from oranges, with the remainder in the peel. KOIVISTOINEN et al. (1964) applied carbamate and organophosphates as dips to harvested fruit and vegetables. The amount of residue removed by simple washing in running water was from zero to 79 percent. The wide variation was both due to type of pesticide and type of fruit or vegetable product used. Care must be taken in sampling procedures used in experiments of this type or variations in results can occur (LYKKEN 1963). WHEELER et al. (1967 a and b) reported the uptake by plants of labelled dieldrin. This group of workers also reported the use of a chloroform-methanol (1:1) extraction was needed quantitatively to remove the pesticide from the tissues. Cotner *et al.* (1968) reported that root-adsorbed dieldrin was present in interior leaf tissues and membranes in wheat plants. This again required severe extraction conditions to recover the labelled dieldrin.

From this you can see the complexity of the problem in studies of this type. First, each insecticide and form of the insecticide used as spray, dust, or emulsion type must be studied. Each fruit or vegetable type and variety is a variable to consider. Weather, soil types, and other atmospheric conditions enter in when plant materials are involved. In animals, method of contamination, diet, and plan of nutrition would effect amount of insecticide located in the final food product.

III. Processing treatments

Most of the research reported on removal of insecticide residues from foods has been from studies using conventional commercial or home processing procedures. Street (1968) referred to this as "adventitious removal" which is incidental to the processing operation. The results reported by Dugan (1968) indicated that this is an important factor in reducing the amounts of residues from levels present on raw agricultural foods as contrasted to levels in foods as consumed. This factor is apparently considered in establishing residue tolerances. Many tolerances were established prior to having the information on processing effects which might indicate study as to validity of those tolerances is in order.

Special processing techniques intentionally to remove all chemical residues have not gained much support. In many cases, industrial concerns do not want to get involved because of cost and the possibility of adverse publicity. So long as an adequate supply of raw agricultural foods is available we can afford this luxury. Someday, as with removal of radioactive materials, it may be necessary to develop techniques to improve adventitious removal or to get complete removal of chemical residues from processed foods. This may need government research support if the work is to be accomplished.

IV. Animal food products

Dugan (1968) indicated this group was a major contributor to chlorinated pesticides in the diet of the U.S. consumer. The animals and chickens can metabolize most of the pesticides other than chlorinated hydrocarbons and therefore these chemicals usually do not appear in foods from animal sources. Chlorinated pesticides are concentrated in fatty tissues and in normal processing are removed when lipid tissues are rendered out, trimmed off, or concentrated into one fraction of the food. The other possibilities are for destruction or residues by heat processing, irradiation, bacterial fermentations, or removal during concentration of foods under vacuum or by high temperature deodorization of milk fat.

The first report by MANN *et al.* (1950) indicated pasteurization of milk had little effect on DDT. DDT was proportional to fat content in all products made from contaminated milk. Since 1964 a number of reports have appeared involving various chlorinated insecticides in processed dairy products. LANGLOIS *et al.* (1964 and 1965) fed lactating dairy cows lindane, DDT, dieldrin, heptachlor, or endrin, and also added these chemicals directly to milk prior to processing the contaminated milks into dairy products using pilot plant or commercial processing equipment. Following this STEMP and LISKA (1966 a) and McCASKEY and LISKA (1967) extended this study to include telodrin, methoxychlor, endosulfan, and chlordane insecticides. Some of the insecticides as heptachlor, DDT, and endosulfan were metabolized by the cow to yield identifiable metabolites.

Following separation of the milk into cream and skimmilk fractions residues were almost entirely in the fat fraction leaving the skimmilk fraction residue free. In this one step, a major fraction of contaminated milk could be salvaged using present commercial processing operations. This could be commercially feasible especially if the nonfat fraction of milk carried more of the dollar value of the product as it may in the future. As indicated in a study by LISKA (1965) processing of the cream fraction into butteroil could be used to remove from 28 to 53 percent of the DDT present. KROGER (1967) and BILLS and SLOAN (1967) showed further processing of butteroil using high temperature vacuum distillation could be a method of essentially removing 100 percent of DDT, dieldrin, and heptachlor epoxide. This procedure is similiar to commercial processing of vegetable oils which reduces residues in finished oils as reported by DUGGAN (1968).

Butter, cheese, and ice cream were manufactured from milk containing selected residues. There was essentially no change in residues on a fat basis in the studies of LANGLOIS *et al.* (1964 and 1965). MONTOURE and MULDOON (1967) reported some change in structure of DDT during cheese processing and ripening. Bacterial fermentations in some cases can effect changes in chemical residues in foods. KIM and HARMON (1967) and BRADLEY and LI (1968) indicated that methoxychlor had no effect on bacterial cultures used in cheese fermentations but that dieldrin caused a reduction in acid production in cheese processing. LEDFORD and CHEN (1968) indicated some degradation of DDT and DDE occurred during growth of cheese microorganisms. Since heat processing is a minimum amount in these products little change due to heat can be expected.

In other experiments, contaminated milk was processed into spray and drum-dried whole milk and sterilized milk. These processes include concentration or removal of water using heat and high vacuum. Also spray or roller drying includes use of high temperatures and volumes of hot air to remove moisture. These conditions are more severe and could be expected to result in some loss of residues. Also, in can sterilization of evaporated milk near a neutral pH is favorable to residue degradation.

The results of some studies at Purdue are summarized in Tables I and II. Dieldrin and heptachlor epoxide residues in milk were

Table I. *Destruction of DDT, lindane, dieldrin, heptachlor epoxide, and heptachlor during the processing of milk into sterilized and dried milk* (Langlois *et al.* 1964 and 1965)

Milk product	Residue (p.p.m. fat basis)				
	DDT	Lindane	Dieldrin	Heptachlor	Heptachlor epoxide
Raw	26.0 [a]	25.0	19.3	3.7	21.2
Condensed	23.1	26.4	11.4	3.9	11.0
Sterilized	25.1	25.6	11.2	3.8	12.6
Spray-dried	9.9	4.4	8.5	0.2	7.4
Drum-dried	4.3	9.3	9.1	0.2	8.1

[a] Average of three trials; averaged to nearest 0.1 p.p.m.

Table II. *Destruction of telodrin, methoxychlor, chlordane, endosulfan, and endosulfan sulfate during the processing of milk into sterilized and dried milk* (Stemp and Liska 1966 a, McCaskey and Liska 1967)

Milk product	Residue (p.p.m. fat basis)				
	Telodrin	Methoxy-chlor	Chlordane	Endosulfan	Endosulfan sulfate
Raw	18.3 [a]	23.1	19.7	15.9	15.2
Condensed	17.8	21.1	17.5	11.4	12.6
Sterilized	10.2	18.6	10.8	9.9	8.8
Sprayed-dried	16.3	22.7	14.7	10.1	8.8
Drum-dried	16.5	23.3	9.9	8.0	4.5

[a] Average of two trials; averaged to nearest 0.1 p.p.m.

reduced from 40 to 50 percent by condensing the milk to one-half the original volume. The rest of the insecticides were essentially unaffected by the treatment. Heat sterilization of the condensed milk caused 20 to 30 percent reduction in telodrin, chlordane, and endosulfan sulfate. Other residues were stable to the heat treatment. Spray drying and drum drying processes resulted in significant decreases in most of the residues studied. The question arises as to whether the residues are destroyed by the heat or are removed with the water during the drying processes. Lindane, one of the more volatile chlorinated pesticides, is reduced more by spray drying than by roller drying which indicates codistillation or removal with water maybe occurring. In drum drying the heat treatment may be sufficient to cause heat destruction of some residues. Drum drying as well as heat sterilization caused degradation of DDT to form TDE (DDD) and DDE. Changes in structure of other residues were not detected. Li and Bradley (1967) irradiated milk using ultraviolet light. They indicated some changes in structure and destruction of methoxychlor and DDT. However, the treatment rendered the milk unacceptable in flavor so has little merit.

An early report (Carter et al. 1948) on effects of meat cooking on DDT residues indicated only frying and pressure cooking removed noticable amounts of residue from beef. This would be expected since little beef fat would be rendered out by other cooking treatments. In contrast, Liska et al. (1967) and McCaskey et al. (1968) reported

Table III. *Chlorinated insecticide residue levels in abdominal fat and fat drippings [a] from hen carcasses containing various chlorinated insecticides* (Liska et al. 1967)

Insecticide	Level administered (p.p.m.)	Sample [b]		
		1	2	3
Lindane	100	5.4 [c]	5.2	5.1
	50	2.1	2.0	2.0
	10	0.8	0.8	0.7
DDT [d]	100	6.1	5.9	6.0
	50	2.2	2.1	2.2
	10	0.9	0.9	0.9
Dieldrin	100	5.2	5.0	5.1
	50	2.2	2.1	2.2
	10	0.9	0.9	0.9
Heptachlor [e]	100	6.7	6.0	6.3
	50	2.3	2.0	2.1
	10	0.9	0.6	0.8
Endrin	100	6.4	6.2	6.0
	50	2.4	2.3	2.2
	10	1.0	0.9	0.9

[a] Three hours at 15 psi.
[b] 1 = abdominal fat, 2 = fat drippings at one hour, and 3 = fat drippings at three hours.
[c] Average of five samples to nearest 0.1 p.p.m.
[d] DDT and breakdown products.
[e] Heptachlor and heptachlor epoxide.

Table IV. *Chlorinated insecticide residues content/g. of fat ^{a} in raw and processed ^{b} tissue samples from hen carcasses containing various chlorinated insecticides* (McCASKEY *et al.* 1968)

Insecticide	Level administered (p.p.m.)	Sample ^{c}			
		1	2	3	4
DDT	0.10-0.15	0.6 ^{d}	3.6	— ^{e}	—
	10-15	7.7	22.5	7.1	17.9
Dieldrin	0.10-0.15	2.5	6.7	3.8	—
	10-15	12.6	26.0	16.6	23.3
Heptachlor	0.10-0.15	9.0	8.9	8.1	2.7
	10-15	6.1	9.1	6.3	6.7
Lindane	0.10-0.15	—	—	—	—
	10-15	0.7	0.9	—	—

^{a} p.p.m. insecticide times 100 divided by percent lipid.

^{b} Three hours at 190° to 200° F.

^{c} 1 = dark meat at zero hour, 2 = white meat at zero hour, 3 = dark meat at three hours, and 4 = white meat at three hours.

^{d} Average of five samples to nearest 0.1 p.p.m.

^{e} Residue level below 0.1 p.p.m. and could not be accurately calculated.

that chicken meat can be heat processed to remove significant amounts of chlorinated pesticide residues. Fat rendered from chicken tissues during processing contained about the same levels of residues as raw lipid tissues. Pesticides used in these studies were DDT, dieldrin, heptachlor, lindane, chlordane, telodrin, and ovex. Contaminated chicken carcasses were simmered in water at 190° to 200° F. for three hours or heat treated at 250° F. under pressure for one or three hours. Simmering removed 40 to 50 percent of the residues. Over 95 percent of all insecticide residues except heptachlor were removed by 250° F. for one hour heat treatments. Heptachlor was the most tightly held of the insecticides studied. Also insecticides were rendered from fat, light, and dark tissues at different rates. It appears that structure of the insecticide may affect the ease of removal from chicken tissue. RITCHEY *et al.* (1967 and 1968) evaluated frying and baking for removal of DDT and lindane from chicken tissues. Removal was in excess of 50 percent and conversion of DDT to DDD occurred at the high temperatures used.

V. Fruits and vegetables

CROSBY (1965) summarized research on the effects of fruit and vegetable processing on insecticide residues. He indicated a need for research in this area. Since that time research studies have been undertaken by a number of research groups. The *National Canners Associa-*

tion (1963 and 1964) has developed radioisotope tracer techniques for the study of removal of pesticide residues from fruits and vegetables. Another report from the *National Canners Association* (1967) contains a rather complete study on effect of processing on residues on vegetables. HEMPHILL *et al.* (1967), FARROW *et al.* (1968), LAMB *et al.* (1968), BALDWIN *et al.* (1968), and CARLIN *et al.* (1966) have also published on processing studies of vegetables and fruits.

Most of the work to date has been on normal processing procedures. However, a recently announced *U.S. Department of Agriculture* supported study by the *National Canners Association* will attempt alterations in food processing to minimize residues in processed foods.

Washing procedures used in processing fruits and vegetables either on an commercial or home scale removed 78 to 90 percent of DDT on tomatoes (*National Canners Association* 1967). BALDWIN *et al.* (1968) reported washing of apples only removed 10 to 15 percent of DDT present. Wax in the apple peel must adsorb the DDT residues since peeling removed another 30 to 40 percent of DDT residues on the apples. The *National Canners Association* (1967) report indicated that use of detergents improved the removal of various residues by washing procedures in potatoes, tomatoes, and spinach. HEMPHILL *et al.* (1967) reported Guthion was easily removed from green beans by washing. These results emphasize effect of type of insecticide, type of food, and washing technique on removal of residues. Possibly a water-charcoal suspension would aid removal of surface residues on fruits and vegetables.

Blanching with hot water or steam aids in removal of DDT, carbaryl, and malathion (*National Canners Association* 1967). CARLIN *et al.* (1966) also reported blanching reduced DDT and Guthion residues on green beans. Peeling and trimming remove considerable amounts of residue not removed by washing treatments (*National Canners Association* 1967, CARLIN *et al.* 1966, BALDWIN *et al.* 1968). CROSBY (1965) further summarizes work on other fruits and vegetables. Cooking of fruits and vegetables may decrease residues depending on temperature used, insecticide involved, and whether cooking is done in contact with metal or glass containers (*National Canners Association* 1967), HEMPHILL *et al.* 1967, CARLIN *et al.* 1966). CROSBY (1965) reported other studies of residue removal by cooking. DDT is converted to DDD in some cooking procedures. Apparently iron or aluminum containers can catalyze the conversion.

The *National Canners Association* reports (1963 and 1964) are important in outlining methods using labelled residues for checking on removal of pesticides in processing operations. In this type of research it is important to be certain that residues analyzed at various steps represent all residues present and not only those easily extracted. WHEELER *et al.* (1967 a and b) reported on research on bound residues in plant tissues which could apply in research summarized above.

Normal processing procedures definitely reduce the level of various residues in man's food supply. This is emphasized by STREET (1969) in the *National Canners Association* (1967) report and implied in the report by DUGGAN (1968). For some pesticides and some foods processing procedures could be varied to remove more residue if the need should arise. Additional research is underway to gain information in this general area.

Summary

Man's food supply at present has a low enough level of insecticide residues in order not to affect human health. Normal processing does contribute to maintainance of acceptable residue levels by causing residue reductions. Removal of residues in foods by processing is affected by type of food, insecticide type and severity of processing procedure used. Variations in present processing procedures or new processing procedures could be developed to remove more residues from foods should the need arise.

Résumé *

Effets de la préparation des aliments sur leur teneur en pesticides

Le niveau des résidus d'insecticides dans l'alimentation actuelle est suffisamment bas pour ne pas affecter la santé de l'homme. La préparation normale des aliments contribue effectivement au maintien de teneurs en résidus acceptables en réduisant leur taux. L'élimination des résidus dans les aliments par la préparation est fonction de la qualité de l'aliment, de celle de l'insecticide et de la sévérité du traitement utilisé. Des changements dans les techniques actuelles ou de nouvelles méthodes de préparation des aliments pourraient être développées pour enlever davantage de résidus des aliments si le besoin s'en faisait sentir.

Zusammenfassung **

Wirkung von Verarbeitungsprozessen auf Pestizide in Nahrungsmitteln

Der gegenwärtige menschliche Nahrungsvorrat enthält Insektizidrückstandsmengen, die niedrig genug sind, um nicht die menschliche Gesundheit zu gefährden. Normale Prozesse tragen zur Erhaltung von annehmbaren Rückstandsmengen bei, indem sie Rückstandsreduzierungen verursachen. Entfernung von Rückständen in Nahrungsmitteln durch Verarbeitung wird durch die Art des Nahrungsmittels, Art des Insektizids und die Härte der angewandten Verarbeitungsprozedur beeinflusst. Veränderungen in den gegenwärtigen Verarbeitungsprozessen oder neue Verarbeitungsprozeduren könnten entwickelt werden,

---

* Traduit par R. MESTRES.
** Übersetzt VON A. SCHUMANN.

um mehr Rückstände von Nahrungsmitteln zu entfernen, falls die Notwendigkeit entsteht.

References

ANDERSON, C. A., D. MACDOUGALL, J. E. KESTERSON, T. HENDRICKSON, and R. F. BROOKS: The effect of processing on Guthion residues in oranges and orange products. J. Agr. Food Chem. 11, 422 (1963).

BALDWIN, R. E., L. K. GONNERMAN, and D. D. HEMPHILL: Pilot study of DDT and its derivatives in apples as affected by preparation procedures. Food Technol. (In press) (1969).

BILLS, D. D., and J. L. SLOAN: Molecular distillation treatment of milk fat containing chlorinated insecticides. J. Dairy Sci. 50, 962 (1967).

BOURNE, M. C.: Annual report of NCM-37. Trace levels of pesticide residues in agricultural commodities in marketing channels. CSRS, U.S. Department of Agriculture (1967).

BRADLEY, R. L., JR., and C. F. LI: The effect of dieldrin on acid development during manufacture of cheddar cheese. J. Milk Food Technol. 31, 202 (1968).

CARLIN, A. F., E. T. HIBBS, and P. A. DAHM: Insecticide residues and sensory evaluation of canned and frozen snap beans field sprayed with Guthion and DDT. Food Technol. 20, 80 (1966).

CARTER, R. H., P. E. HUBANKS, and H. D. MANN: Effect of cooking on the DDT content of beef. Science 107, 347 (1948).

COTNER, R. C., R. H. HAMILTON, R. O. MUMMA, and D. E. H. FREAR: Localization of dieldrin in wheat tissue. J. Agr. Food Chem. 16, 608 (1968).

CROSBY, D. G.: The intentional removal of pesticide residues. In C. O. CHICHESTER (ed.): Research in pesticides, p. 213. New York: Academic Press (1965).

DUGGAN, R. E.: Residues in food and feed. Pesticides Monitoring J. 1, 2 (1968).

——, and K. DAWSON: A report on pesticides in food. F.D.A. Papers 1, 4 (1967).

DURHAM, W. F.: Pesticide residues in foods in relation to human health. Residue Reviews 4, 63 (1963).

FARROW, R. P., F. C. LAMB, R. W. COOK, J. R. KIMBALL, and E. R. ELKINS: Removal of DDT, malathion, and carbaryl from tomatoes by commercial and home preparative methods. J. Agr. Food Chem. 16, 65 (1968).

Federal Register 32, 4059 (1967).

HEMPHILL, D. D., R. E. BALDWIN, A. DEGUZMAN, and H. K. DELOACH: Effects of washing, trimming and cooking on levels of DDT and derivatives in green beans. J. Agr. Food Chem. 15, 290 (1967).

KIM, S. C., and L. G. HARMEN: Effect of pesticide residues on growth and fermentation ability of Streptococcus lactis and Lactobacillus casei. J. Dairy Sci. 50, 939 (1967).

KOIVISTOINEN, P., M. KONONEN, A. KARINPAA, and P. ROINE: Stability of malathion residues in food processing and storage. J. Agr. Food Chem. 12, 557 (1964).

KROGER, M.: Removal of certain organochlorine hydrocarbon pesticides from milk fat by steam deodorization. J. Dairy Sci. 50, 944 (1967).

LAMB, F. C., R. P. FARROW, E. R. ELKINS, R. W. COOK, and J. R. KIMBALL: Behavior of DDT in potatoes during commercial and home preparations. J. Agr. Food Chem. 16, 272 (1968).

LANGLOIS, B. E., B. J. LISKA, and D. L. HILL: The effects of processing and storage of dairy products on chlorinated insecticide residues. I. DDT and lindane. J. Milk Food Technol. 27, 264 (1964).

—— —— —— The effects of processing and storage of dairy products on chlorinated insecticide residue. II. Endrin, dieldrin and heptachlor. J. Milk Food Technol. 28, 9 (1965).

LEDFORD, R. A., and J. H. CHEN: Degradation of DDT and DDE by cheese microorganisms. Abstr. 28th Annual Meeting Inst. Food Technol. (1968).

LI, C. F., and R. L. BRADLEY, JR.: Degradation of chlorinated hydrocarbons in fluid milk systems by ultra-violet light. J. Dairy Sci. 50, 944 (1967).

LICHENSTEIN, E. P., G. R. MYRDAL, and K. R. SCHULA: Absorption of insecticidal residues from contaminated soils into five carrot varieties. J. Agr. Food Chem. 13, 126 (1965).

LISKA, B. J.: Effects of processing on residues in foods. Proc. 25th Semi-Annual Meeting Amer. Feed Manufactures Assoc. Nutrition Council, p. 35 (1965).

—— Effects of processing on pesticide residues in milk. J. Animal Sci. 27, 827 (1968).

——, A. R. STEMP, and W. J. STADELMAN: Effect of method of cooking on chlorinated insecticide residue content in edible chicken tissues. Food Technol. 21, 117A (1967).

LYKKEN, L.: Important considerations in collecting and preparing crop samples for residue analysis. Residue Reviews 3, 19 (1963).

MANN, H. D., R. H. CARTER, and R. E. ELY: The DDT content of milk products. J. Milk Food Technol. 13, 340 (1950).

MARTH, E. H.: Residues and some effects of chlorinated hydrocarbon insecticides in biological material. Residue Reviews 9, 1 (1965).

McCASKEY, T. A., and B. J. LISKA: Effect of milk processing on endosulfan, endosulfan sulfate and chlordane residues in milk. J. Dairy Sci. 50, 1991 (1967).

——, A. R. STEMP, B. J. LISKA, and W. J. STADELMAN: Residue in egg yolks and raw and cooked tissues from laying hens administered selected chlorinated hydrocarbon insecticides. Poultry Sci. (In press) (1969).

MONTOURE, J. E., and P. J. MULDOON: Effect of manufacturing and storage on the chlorinated hydrocarbon residue level in Monterey and cheddar cheese manufactured from DDE, TDE, DDT contaminated milk. J. Dairy Sci. 50, 954 (1967).

National Canners Association: Radioisotopic tracer techniques for evaluation and improvement of industry practices in removal of pesticide residues from food. SAN-1022 rept., Washington, D.C. (1963).

—— Radioisotopic tracer techniques in evaluation and improvement of industry practices for removal of pesticide residues from foods. SAN-536-10 rept., Washington, D.C. (1964).

—— Investigations on the effect of preparation and cooking on the pesticide residue content of selected vegetables. Final rept., Washington, D.C. (1967).

RITCHEY, S. J., R. W. YOUNG, and E. O. ESSARY: The effects of cooking on chlorinated hydrocarbon pesticide residues in chicken tissues. J. Food Sci. 32, 238 (1967).

—— —— —— Effects of cooking methods and heating on DDT in chicken tissues. Abstr. 28th Annual Meeting, Inst. Food Technol. (1968).

STEMP, A. R., and B. J. LISKA: Effects of processing and storage of dairy products on telodrin and methoxychlor residues. J. Dairy Sci. 48, 1006 (1966).

STINE, C. M., and S. PATTON: Preparation of milk fat. II. A new method of manufacturing butteroil. J. Dairy Sci. 35, 655 (1952).

STREET, J. C.: Methods of removal of pesticide residues. J. Can. Med. Assoc. (In press) (1969).

WHEELER, W. B., D. E. H. FREAR, R. O. MUMMA, R. H. HAMILTON, and R. C. COTNER: Quantitative extraction of root-absorbed dieldrin from the aerial parts of forage crops. J. Agr. Food Chem. 15, 227 (1967 a).

—— —— —— —— —— Absorption and translocation of dieldrin by forage crops. J. Agr. Food Chem. 15, 231 (1967 b).

Canning operations that reduce insecticide levels in prepared foods and in solid food wastes

By

R. P. Farrow,* E. R. Elkins,* W. W. Rose,** F. C. Lamb,**
J. W. Ralls,** and W. A. Mercer **

Contents

I. Introduction

Pesticide residue tolerances established by the *U.S. Food and Drug Administration* apply to the raw agricultural commodity. Canned or frozen food products are not specifically mentioned in the detailed regulations listing the tolerances. The law simply provides that no action will be taken against a processed food if the residues are within tolerance and if the pesticide compounds have been removed to the extent possible in good manufacturing practice.

The intent of the regulatory agencies and food manufacturers, however, is to minimize the exposure of the public to pesticide compounds whenever possible, even though there is no question of safety involved in the traces of pesticide compounds that remain at harvest. This philosophy prompts an active interest in research on the extent to which pesticide residues can be removed during the washing and canning operations that intervene between harvest and use. A reason-

---

* National Canners Association, Washington, D.C.
** National Canners Association, Berkeley, California.

able concern over the fate of pesticide compounds in our environment also prompts an active interest in the ultimate disposition of those portions of the residue that are removed by washing and processing operations.

II. Canning operations

During the last four years the research laboratories of the *National Canners Association* have studied canning operations to determine their effectiveness in removing or destroying the traces of pesticide compounds that may remain as legal residues on raw agricultural commodities. Let us quickly review some of the "unit operations" involved in a modern canning plant and comment on their potential effects on the pesticides applied during the growing season (Table I).

Table I. *Canning plant "unit operations"*

Harvesting, husking, shelling
Washing
Blanching
Peeling, juice extraction
Filling and closing
Processing
Disposal of liquid wastes
Disposal of solid wastes

The recovery of the edible portion of the vegetable or fruit may involve husking, peeling, or shelling operations which effectively remove all of the pesticide compound with the discarded portions of the plant. Pesticide compounds seldom if ever come into direct contact with the edible portion of sweet corn, for example, since they are applied to exterior portions removed in the harvesting operation. Similar considerations apply to green peas which are commercially harvested on the vine and separated mechanically from the vines and pods.

On arrival at the canning plant the product is subjected to several immersion or spray washing operations which sometimes involve the use of a detergent to aid in the cleaning. Blanching is a short treatment in hot water or steam applied to most vegetables such as peas, green beans, spinach, and broccoli. Washing and blanching may remove much of the pesticide residue. Most fruits are subjected to peeling or juice extraction. The peel and extracted plant material constitute a major portion of the solid waste resulting from procedures applied within the canning plant itself.

Filling and closing are of no significance in connection with the pesticide residue content, except for the diluting effect of the added sirup or brine.

The heat process itself is ordinarily carried out on the closed and sealed container, and could be expected to result in substantial destruction of compounds subject to hydrolysis or heat effects. After processing the canned product is cooled, labeled, and cased for shipping.

Liquid wastes result from the washing, blanching, and lye peeling operations. In some instances the liquid wastes may contain much of the pesticide compound that was originally on the product. It would be greatly diluted, however, and the method of disposal would not be expected to return significant quantities of pesticide to areas supporting crops or fish and game.

Solid waste materials may result from juice extraction, trimming, cleaning, and mechanical peeling operations. In those instances in which the pesticide residue is confined largely to the skin of the fruit, the solid waste may have a considerably higher pesticide content than that of the raw fruit entering the canning plant. Of course, the absolute amounts of pesticide are quite small in comparison with the quantities originally applied to the growing plant. These waste materials are frequently disposed of by means of a sanitary land fill operation.

III. Experimental methods

The data available on the effects of these canning operations on the pesticide compounds in Table II result from three research projects.

Table II. *Pesticide compounds and crops studied*

DDT		Carbaryl
Tomatoes		Tomatoes
Spinach		Green beans
Green beans		Spinach
Potatoes		Broccoli

Malathion	Parathion	Diazinon
Tomatoes	Spinach	Tomatoes
Green beans	Broccoli	Green beans
		Spinach

A *U.S. Public Health Service* (PHS) research grant EF-00611-03 from the Division of Environmental Engineering and Food Protection enabled us to study the residue levels and transformation products of diazinon on several products. A research contract from the *U.S. Department of Agriculture* Agricultural Research Service resulted in data on the removal of DDT, malathion, parathion, and carbaryl during canning operations. Equipment obtained as a result of the *U.S. Atomic Energy Commission* Contract AT(04-3)-536 was used in the washing studies. Studies on the effect of composting operations on various pesticides were supported in part by PHS research grant UI-00553 from National Center for Urban and Industrial Health.

IV. Pesticide removal by washing

The extent to which pesticide residues may be removed by commercial cannery washing operations is influenced by a variety of factors. Included among these are the chemical properties of the pesticide compound, the nature of the food commodity, the length of time that the residue has been in contact with the food, and the formulation in which the pesticide has been applied. These factors should be kept in mind in attempting to generalize on the results of pesticide residue removal studies. The results we will present have been selected as representative of those obtained during several years of research in our laboratories on this subject.

When applied under the conditions used in our work as a 50 percent wettable powder, the major portion of the DDT (Fig. 1) on

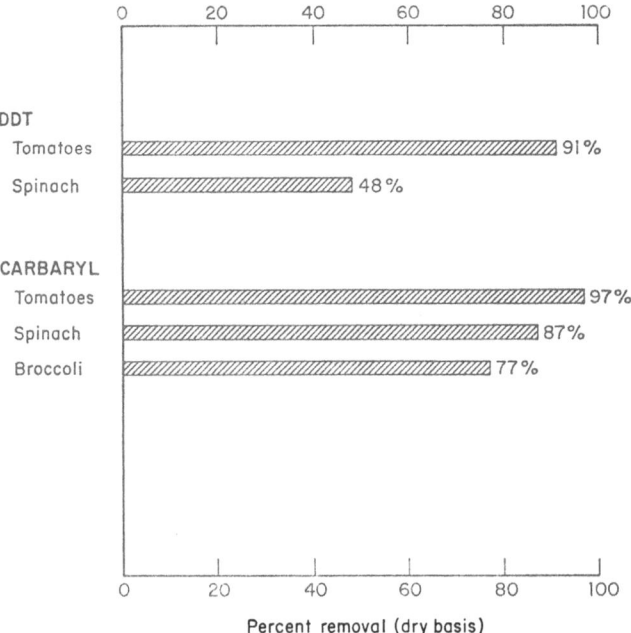

Percent removal (dry basis)

Fig. 1. Pesticide removal by water washing: DDT and carbaryl

tomatoes is apparently present as a loosely held surface residue easily removed with a simple water wash. As one might expect, it is somewhat more difficult to remove the same type of formulation from spinach. About half of the DDT residue was removed by a water wash. Carbaryl, in the form used in our work, was easily removed from tomatoes, spinach, and broccoli (Fig. 1.). Its removal from tomatoes is essentially complete while major portions, 87 and 77 percent, respectively, were removed from spinach and broccoli.

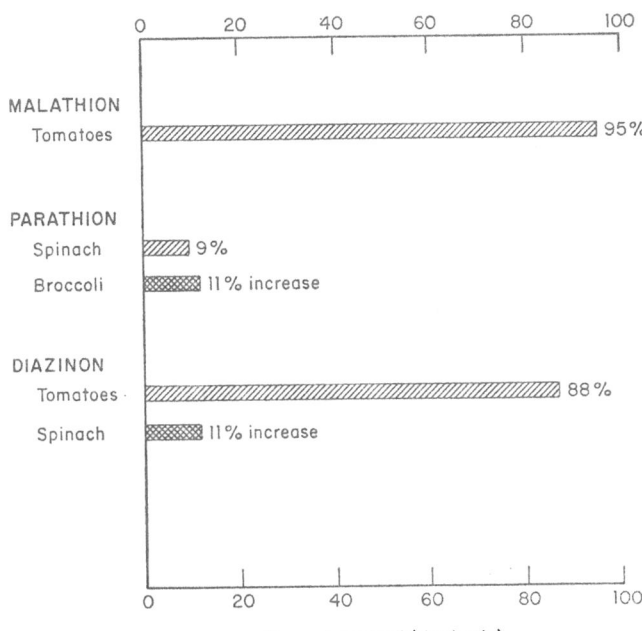

Fig. 2. Pesticide removal by water washing: malathion, parathion, and diazinon

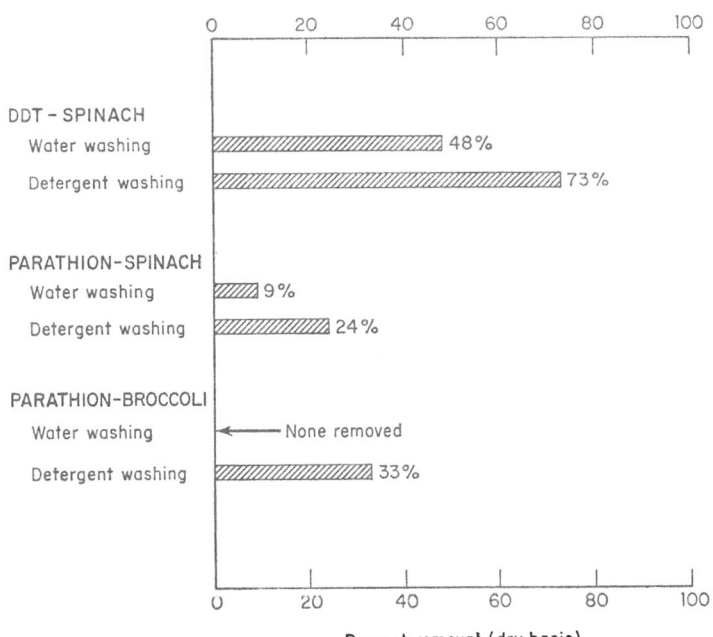

Fig. 3. Detergents as aids in pesticide removal: DDT-spinach, parathion-spinach, and parathion-broccoli

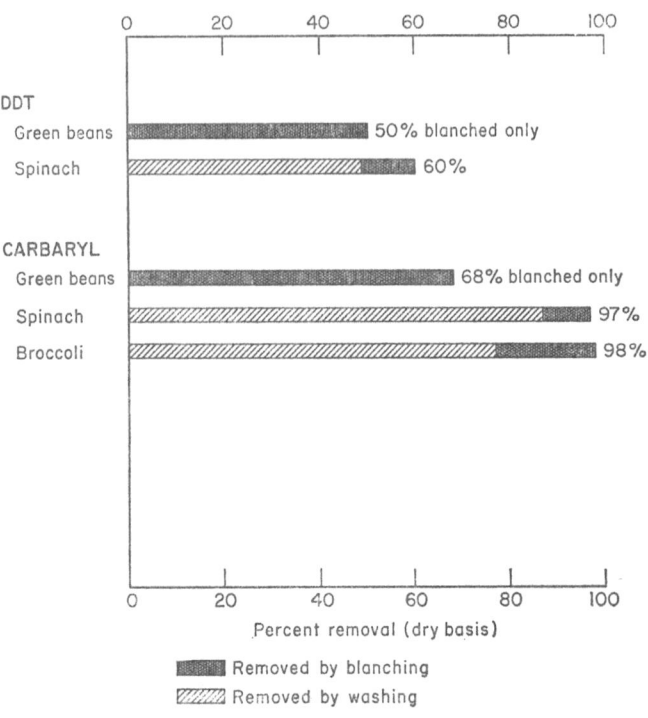

Fig. 4. Pesticide removal by washing plus blanching: DDT and carbaryl

The organophosphates studied were malathion, parathion and diazinon (Fig. 2). Malathion was easily removed from tomatoes by washing. All the pesticide formulations used in our study were removed easily from tomatoes. We were somewhat surprised at the tenacity of the parathion residues. Washing operations had little effect on them. We obtained an apparent removal of about nine percent of the parathion from spinach and an apparent increase of about 11 percent of the parathion residues in broccoli. Both spinach and broccoli have large surface areas and are therefore subject to considerable leaching of soluble solids during washing operations. Since results are expressed on a dry weight basis, this can lead in some instances to an apparent increase in the residue and tends to lower the "percent removal" results expressed on a dry basis. A similar situation was encountered with diazinon residues in spinach.

Several detergents have been cleared for use as aids in washing products in preparation for canning (Fig. 3). For a few pesticide product combinations, detergents increase the proportion removed by washing operations. Detergents significantly improved removal of DDT residues from spinach, and were beneficial in increasing removal of parathion from spinach and broccoli.

V. Removal by washing and blanching

Commercial blanching operations are carried out in either hot water or steam. The hot water blanch is probably more commonly used in commercial canning operations and the results in Figure 4 were obtained with water blanches. As would be expected, it is more effective in removing pesticide residues than steam blanching. In the commercial canning of green beans, the beans are blanched and washed in the same operation. Our experimental procedure followed commercial practice in this respect. We obtained about 50 percent removal of the DDT residue and about 68 percent of the carbaryl residue. About half of the DDT residue in spinach was removed in the wash. An additional 10 percent was removed by the much more severe treatment provided by a hot water blanch. Evidently the washing removes the loosely held surface material. The remaining portion is more difficult to remove. Washing had removed the major portion of the carbaryl residues from spinach and broccoli. During the blanching operation, the removal of this residue was virtually completed.

With one exception (Fig. 5), blanching was effective in removing

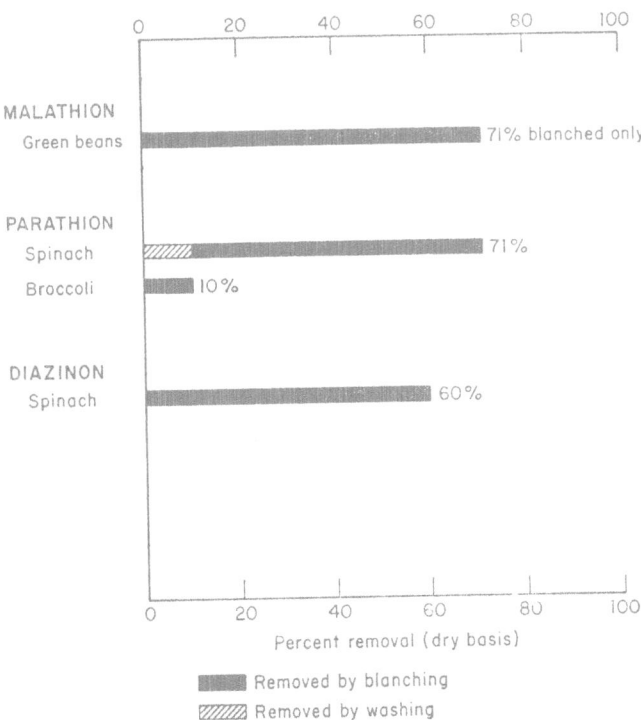

Fig. 5. Pesticide removal by washing plus blanching: malathion, parathion, and diazinon

substantial portions of the organophosphate compounds. This canning step removed over 70 percent of the malathion from green beans. Parathion on spinach was little affected by the washing operation. Hot water blanching, however, removed a considerable portion. It was not effective in removing parathion from broccoli. These results are affected by the leaching of soluble solids, making the blanching appear less effective than it actually is. Although washing was not effective in removing diazinon from spinach, the hot water blanch removed 60 percent.

VI. Removal by washing plus peeling

Nearly all fruits and root crops are subjected to chemical or mechanical peeling operations prior to canning. It is almost universally true that where peeling operations are used, the pesticide removal is virtually complete. DDT, malathion and carbaryl were all completely removed from tomatoes by washing and peeling, as shown in Figure 6.

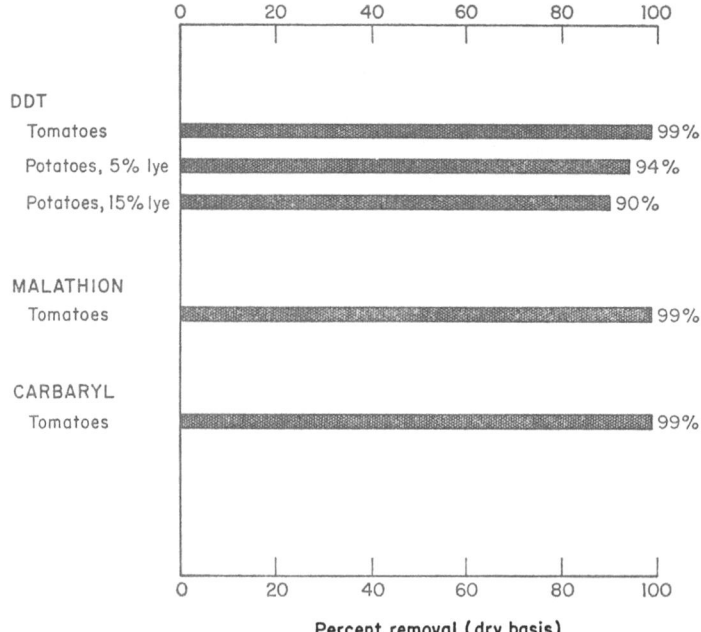

Fig. 6. Pesticide removal by washing plus peeling: DDT, malathion, and carbaryl

DDT residues in potatoes, resulting from its presence in the soil, were removed to the extent of 90 percent or more by washing and peeling. These results are in agreement with industry experience gained from the monitoring of root crops. Carrots, for example, may take up traces of chlorinated hydrocarbons when grown in soils where these are present. These traces of residue remain almost entirely within the surface area and are removed by the abrasive peeling operation ordinarily used on this crop.

VII. Removal by all canning operations

The total amount of pesticide removed by all of the combined canning operations is shown in Figures 7 and 8. Removal is better

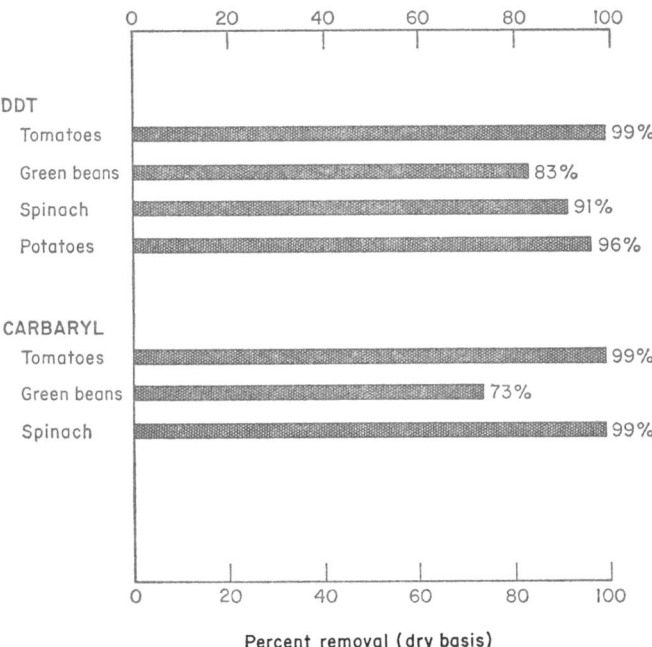

Percent removal (dry basis)

Fig. 7. Pesticide removal by washing plus blanching plus canning: DDT and carbaryl

than 90 percent for DDT on tomatoes, spinach, and potatoes, and for carbaryl on tomatoes and spinach. The green beans retained small portions of DDT and carbaryl residues.

It is a good general rule that when the quantity of pesticide residue is large, greater percentages will be removed in preparation for serving. This observation has been made several times in the course of our experimental work, and also has a result of monitoring activities. Our field application schedules were aimed at achieving a residue at harvest time that would be close to, or slightly in excess of the permissible *Food and Drug Administration* tolerance. To obtain these levels we sometimes made pesticide applications closer to harvest than the minimum time provided in the label registration. Generally speaking our highest percent removal was obtained in such instances. The

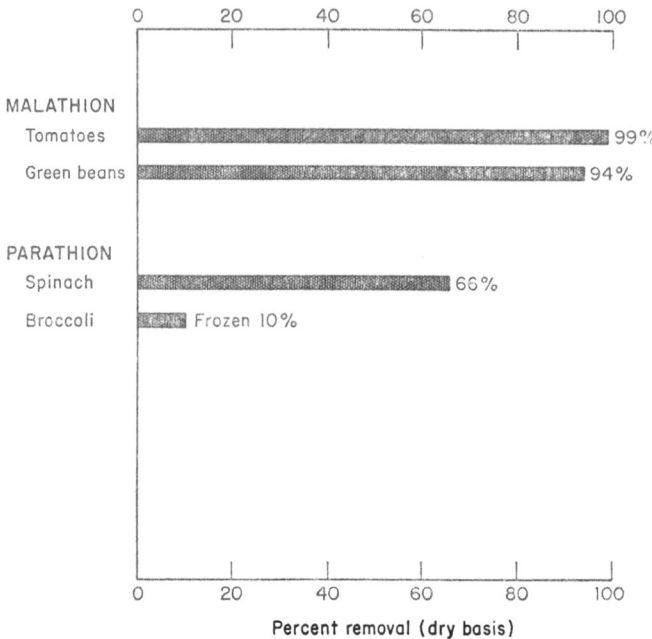

Fig. 8. Pesticide removal by washing plus blanching plus canning: malathion and parathion

DDT application to green beans was made in accordance with label recommendations. The residue at harvest time was about four p.p.m., compared with eight p.p.m. on tomatoes, less than one p.p.m. on potatoes, and about 27 p.p.m. on spinach.

Over-all removal of the organophosphate compounds was very good, with the exception of broccoli. Note (Fig. 8) that the results on this vegetable refer to the frozen product. Broccoli is not canned commercially. The removal of malathion was essentially complete in both tomatoes and green beans. About 66 percent of the parathion was removed from the spinach. On a percentage basis, parathion proved to be the most difficult to remove. It should be pointed out, however, that the absolute amounts remaining are quite small. The initial residues were about 1.5 p.p.m. in spinach and 0.65 p.p.m. in broccoli. After canning, 0.3 p.p.m. of parathion remained in the spinach, on a wet basis, and about 0.4 p.p.m. of parathion on frozen broccoli.

VIII. Pesticides in canning wastes

Pesticide compounds that are tightly held by the fruit or vegetable product are usually located in the peel. This is ordinarily removed as a part of the preparation for canning and may constitute a major portion of the waste material generated by some canning operations. In such instances the pesticide may be concentrated in the solid wastes. The data shown in Table III on the residue content of tomatoes and

Table III. *Residue contents of tomatoes and tomato waste*

Product	Residue (p.p.m.)		
	DDT	Malathion	Carbaryl
Unwashed	7.7	15.9	5.2
Washed	1.15	0.8	0.14
Peeled	Trace	0.1	Trace
Waste	8.2	5.3	0.1

tomato wastes indicate the kind of distribution that may occur. The raw tomatoes in this work contained about eight p.p.m. of DDT, which was almost completely removed by washing and peeling. Nearly all of the pesticide remaining after the wash was located in the peel and, as a result, the waste material had a DDT concentration of about eight p.p.m. A similar situation was found to exist with malathion residues. Carbaryl is a fairly polar compound easily removed by the washing operation. It does not tend to accumulate in the waste material.

Solid canning wastes are disposed of in a variety of ways. Probably the most common disposal procedure is sanitary land fill operations, in which pesticides would be subject to degradative processes similar to those taking place in the soil. Solid food wastes are also sometimes used as animal feed supplements. The presence of pesticide residues may restrict or prevent such uses.

Composting has been under study in our laboratory as an alternative means of disposing of solid cannery wastes. Two types of composting operations have been considered, a batch-type and a continuous thermophilic procedure. The effect of these operations on the four pesticides diazinon, parathion, dieldrin, and DDT was observed throughout a six-month period.

The effect of both composting operations on dieldrin is illustrated in Figure 9. The anomalous results obtained in the early samples may have been due to the fact that the insecticide was not uniformly distributed at the beginning of the process, although the same method of mixing was used in all instances. Dieldrin was reduced in the batch process from an initial concentration of 2.9 p.p.m. to a final residue of 0.2 p.p.m.; the thermophilic process reduced the final concentration to 1.5 p.p.m.

DDT was studied in a similar manner (Fig. 10). The results indicate that in the batch-type composting procedure the DDT was

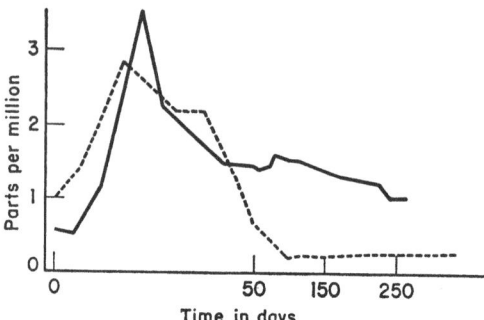

Fig. 9. Decline of dieldrin in concentration during compost process: —— thermophilic procedure, ... batch procedure

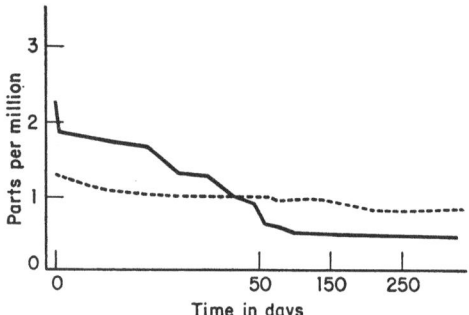

Fig. 10. Decline of p,p'-DDT concentration during compost process: —— thermophilic procedure, ... batch procedure

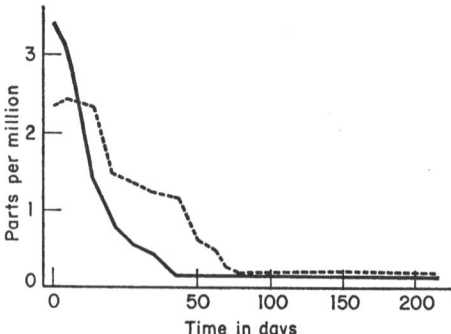

Fig. 11. Decline of parathion concentration during compost process: —— thermophilic procedure, ... batch procedure

relatively resistant to degradation. Analyses performed on the thermophilic samples indicated a steady decrease while high temperatures existed in the compost mass. After the temperature declines, there is no further loss of this insecticide. Known breakdown products such as DDE and TDE, were not detected in any of these samples. The stability of DDT may have been due to aerobic conditions in the compost mass. Recent work by others indicates that it can be dechlorinated by bacteria under anaerobic conditions.

Parathion was degraded rapidly in both the batch and thermophilic processes (Fig. 11). Continuous thermophilic composting reduced the initial level of 3.3 p.p.m. to essentially zero in about 42 days. The batch-type process was also effective in reducing the insecticide concentration but at a somewhat slower rate. Diazinon was also rapidly degraded in both the batch and thermophilic composting methods (Fig. 12). In approximately 10 days the thermophilic temperatures

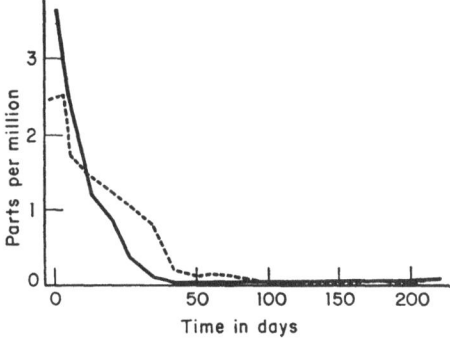

Fig. 12. Decline of diazinon concentration during compost process: ——— thermophilic procedure, ... batch procedure

had reduced the concentration of diazinon by 50 percent. About 28 days of the batch-type composting was required to give the same reduction.

IX. Conclusion

Several years of work on the effect of food preparative operations on pesticide residues have demonstrated that canning plant operations remove most of the pesticide residues present on the crops received for canning. The percent removal varies considerably, depending upon the nature of the canning crop, the pesticide, its formulation, and its weathering history. Generally speaking, the larger the residue the greater the percent removal.

Loosely held surface residues are easily removed by washing and blanching operations. Residues which have penetrated into the surface are difficult to remove from green leafy vegetables. Those that penetrate into the surface layers of fruits and root crops are removed in commercial peeling operations with little or no transfer of the residue to the edible portions of the food.

The peels and waste residual material from extraction operations often constitute a major portion of the solid wastes resulting from canning. These are frequently disposed of in sanitary land fills where the pesticide content of the wastes could not be expected to cause adverse effects.

Composting of solid food wastes sharply reduces the pesticide residue content of the material in many instances.

Summary

Commercial canning operations often substantially reduce the level of pesticide residues remaining on the finished food product. Washing removes loose surface residues and major portions of polar compounds, such as carbaryl. Hot water blanching increases pesticide removal and may hydrolyze substantial fractions of non-persistant compounds; steam blanching is less effective. Non-polar pesticides are frequently held tenaciously in the waxy layers of the peel of fruits and vegetables. Peeling and juice extraction operations usually result in almost complete removal of DDT and other chlorinated hydrocarbons. The pesticides remain in the solid waste resulting from these procedures. Thermal processing necessary to preserve low-acid foods results in the partial destruction of DDT and probably malathion. Carbaryl and parathion are relatively heat stable. Composting appears to be feasible method for disposal of some solid wastes, such as those resulting from peeling and screening operations. Diazinon and parathion in solid wastes are degraded fairly rapidly and almost completely. Dieldrin is substantially reduced by some composting procedures, while DDT is resistant to decomposition.

Résumé *

Opérations de conserverie qui réduisent les taux d'insecticides dans les aliments préparés et dans les déchets solides des aliments

Les opérations de mise en conserve réduisent souvent de manière substantielle la teneur des résidus de pesticides subsistant sur les aliments préparés. Le lavage enlève les résidus superficiels et la plus grande partie des composés polaires comme le carbaryl. Le blanchiment à l'eau chaude augmente l'élimination et peut hydrolyser des

---

* Traduit par R. MESTRES.

fractions substantielles de composés non persistants ; le blanchiment à la vapeur est moins efficace. Les pesticides non polaires sont souvent retenus avec tenacité dans les couches cireuses de la peau des fruits et des légumes. L'épluchage et les opérations d'expression des jus conduisent à une élimination à peu près complète du zeidane et des autres hydrocarbures chlorés. Les pesticides demeurent dans les déchets solides donnés par ces opérations. La pasteurisation nécessaire pour stabiliser les aliments à faible acidité produit une destruction partielle du zeidane et probablement du malathion. Le carbaryl et le parathion sont relativement stables à la chaleur. La transformation en engrais paraît être une méthode possible pour l'utilisation de certains déchets solides, tels que ceux résultants des opérations d'épluchage et de triage. Le diazinon et le parathion sont assez rapidement et presque entièrement dégradés dans les déchets solides. L'HEOD est réduit substantiellement par certains procédés de fabrication d'engrais alors que le zeidane résiste à la dégradation.

Zusammenfassung *

Eindosungsverfahren zur Reduzierung von Insektizidmengen in zubereiteten Nahrungsmitteln und in festen Nahrungsmittelabfällen

Kommerzielle Eindosungsverfahren reduzieren oft wesentlich die Menge an Pestizidrückständen, welche im fertigen Nahrungsprodukt verbleiben. Waschen entfernt lose Oberflächenrückstände und grosse Mengen von polaren Verbindungen, wie z. B. Carbaryl. Blanchieren in heissem Wasser steigert die Pestizidentfernung und kann wesentliche Fraktionen von nichtpersistenten Verbindungen hydrolisieren, blanchieren in Dampf ist weniger wirkungsvoll. Unpolare Pestizide werden oft hartnäckig in den Wachsschichten von Fruchtschalen und Gemüsen festgehalten. Schälen und Saftextraktionsverfahren haben für gewöhnlich die fast vollständige Entfernung von DDT und andern chlorierten Kohlenwasserstoffen zur Folge. Die Pestizide verbleiben dann im festen Abfall, der bei diesen Verfahren entsteht. Hitzeverfahren, die notwendig sind, um Nahrungsmittel mit niedrigem Säuregehalt zu konservieren, resultieren in teilweiser Zerstörung von DDT und wahrscheinlich Malathion. Carbaryl und Parathion sind relativ hitzebeständig. Kompostieren scheint eine passende Methode zur Beseitigung von festen Abfällen zu sein wie solchen, die von Schälen und Aussortierungsmethoden herrühren. Diazinon und Parathion in festen Abfällen bauen ziemlich schnell und fast vollständig ab. Dieldrin wird wesentlich durch einige Kompostierungsverfahren reduziert, während DDT abbaubeständig ist.

---

* Übersetzt von A. SCHUMANN.

Chemical and thermal methods for disposal of pesticides

By

M. V. KENNEDY,* B. J. STOJANOVIC,* AND F. L. SHUMAN, JR.*

Contents

I. Introduction

The disposal of pesticide wastes and pesticide containers is one of the pressing problems of modern agricultural leaders. It is of grave concern to many leaders in other segments of the population as well. The problem is merely magnified by the all too frequent reports of accidental poisoning involving these potent chemicals.

Several methods of disposal of these waste pesticides and the resulting containers have been investigated, but none have proven to be ideal procedures although some have varying degrees of merit. Perhaps the most widely circulated recommendation at present is to bury the materials in a sandy soil at a depth of 18 inches in a private disposal site or take them to a sanitary land-fill type of dump (*U.S. Department*

---
* Contribution from the *Mississippi Agricultural Experiment Station,* Mississippi State University, State College, Mississippi. This study was supported by Agricultural Research Service, U. S. Department of Agriculture, Grant No. 12-14-100-9182(34), administered by the Crops Research Division, Beltsville, Maryland.

of Agriculture, ARS, August 1964) (U.S. Department of Agriculture, ARS, New York State College of Agriculture). Of course, this site then cannot be used for agricultural purposes again at any time in the forseeable future.

Another method of disposal which has been used quite extensively is incineration. Here again some serious objections must be overcome if this procedure is to be accepted on a nation-wide basis. Simple incineration without entrainment of the resulting gases would present a definite threat from the standpoint of air polution. Also, it would endanger humans, animals, and vegetation for some distance around the incineration site upon combustion of certain of the pesticides. These, as well as economic problems concerned with the energy source for incineration, need further investigation.

II. Materials studied

Twenty pesticide chemicals were used for the present investigation. Basically the study was divided into two distinct phases. These were thermal treatment of pesticides and chemical treatment of pesticides.

a) Reference standards

2,4-D (isooctyl ester of 2,4-dichlorophenoxyacetic acid)
Picloram (potassium salt of 4-amino-3,5,6-trichloro-picolinic acid)
Atrazine (2-chloro-4-ethylamino-6-isopropylamino-s-triazine
Diuron [3-(3,4-dichlorophenyl)-1,1-dimethylurea]
Trifluralin (a,a,a-trifluoro-2,6-dinitro-N,N-dipropyl-p-toluidine)
Bromacil (5-bromo-3-sec-butyl-6-methyluracil)
DSMA (disodium methanearsonate)
DNBP (alkanolamine salt of 4,6-dinitro-o-sec-butyl-phenol)
Dicamba (2-methoxy-3,6-dichlorobenzoic acid)
Dalapon (sodium salt of 2,2-dichloropropionic acid)
Paraquat (1,1-dimethyl-4,4-bipyridylium salt)
Vernolate (S-propyl dipropylthiocarbamate)
2,4,5-T (trichlorophenoxyacetic acid)
Carbaryl (1-napthyl N-methyl carbamate)
DDT (dichlorodiphenyl-trichloroethane)
Dieldrin (hexachloroepoxyoctahydro-endo, exo-di-
    methanonaphthalene)
Malathion (0,0-dimethyl dithiophosphate of diethyl
    mercaptosuccinate)
PMA (phenyl mercuric acetate)
Zineb (zinc ethylenebisdithiocarbamate)
Nemagon (1,2-dibromo-3-chloropropane)

b) Commercial formulations

2,4-D (4 lb./gal. "Formula 40")
Picloram (11.6% solution)
Atrazine (80% wettable powder)
Diuron (80% wettable powder)
Trifluralin (4 lb./gal. — liquid)
Bromacil (80% wettable powder)
DSMA (3.2 lb./gal.)
DNBP (3 lb./gal. "premerge")
Dicamba (4 lb./gal.)
Dalapon (85% wettable powder)
Paraquat (2 lb./gal.)
Vernolate (6 lb./gal., liquid)
2,4,5-T (4 lb./gal.; 44.1% acid equivalent)
Carbaryl (10% dust)
DDT (technical flakes)
Dieldrin (17.8% solution)
Malathion (5 lb./gal.; 57% solution)
PMA (Mersolite — 88 W; 95% water dispersable)
Zineb ("Parazate" –C; 75% wettable powder)
Nemagon (8.6 lb./gal.)

III. Thermal treatments

Investigations involving thermal treatment or incineration involved three methods: (1) differential thermal analysis, (2) dry combustion, and (3) ashing in a muffle furnace. Additionally, pyrolysis will be used in future investigations to identify the volatile products of incineration.

a) Differential thermal analysis

This method is used primarily for the classification and identification of clays, although recently it has been further developed and has been employed extensively in solving problems of general chemical interest (CHESTERS et al. 1959).

Thermal analysis is defined as "the process of determining the temperature at which changes in atomic rearrangement accompanied by a usually abrupt change in heat content occur in materials, by observing the aberrations which the heat content changes impose on the rate of temperature change of a specimen situated in an environment whose temperature is changing in a known manner" (SMOTHERS and CHANG 1966). Essentially, the analysis consists of recording the temperature of a sample as it is heated and plotting the resulting data. The differential method is considered to be superior to other methods, such as the inverse-rate method or the time-temperature method, because it is less subject to errors due to drafts, power changes, etc.

The differential analysis method employs a test sample and a "dummy" or neutral body as a reference standard.

Both the sample and the standard are heated simultaneously at the same rate through the same temperature interval. When the test sample undergoes rearrangement, the heat effect creates a difference in temperature between the specimen and the reference standard. This difference in temperature is plotted against the temperature at which this difference occurs; however, the temperature is determined by a second thermocouple located in the test assembly.

The thermal effects which take place may be either exothermic or endothermic. Causes of these changes may be boiling, vaporization, sublimation, fusion, or crystalline-structure inversion. Other changes may take place which are primarily chemical, such as oxidation or reduction, dehydration, combination, and decomposition. Many of these changes produce endothermic heat effects; however, oxidation, crystalline-structure inversion, and decomposition may cause exothermic heat effects.

Each compound will give an individually distinctive reproducible curve which can be used to qualitatively identify the substance. Each of the various endothermic and exothermic peaks will have the same shape, number, and position from one analysis to the next.

The curves are not as strictly reproducible as the X-ray diffraction pattern or the infrared spectrum, but experience has shown that careful reproduction of all conditions will give an identical pattern from one analysis to the next.

Differential thermal analyses of pesticides were carried out on a "Deltatherm," Model D2000 [1] differential thermal analysis apparatus. The sample holder block was fabricated from inconel metal, having service for four sample cups, four reference cups, and one temperature monitor cup. The cups were constructed from an alloy of platinum with 13 percent rhodium and had an inside diameter of 3/16 inch. The thermocouples were spot-welded directly to the bottom of the pans.

The material employed in the temperature cups and in the furnace monitor cups was aluminum oxide. In these analyses the temperature of the sample block was increased at a linear rate of 10° C./minute. The recording of the differential thermal curves was accomplished by an electrical current between a scanning stylus and sensitized paper.

In this investigation, no attempt was made to quantify the results obtained, since the primary interest was to establish the temperature for complete combustion. However, the same quantity of sample was used for analysis from one compound to the next.

The temperatures at which endothermic and exothermic peaks and the complete combustion of pesticides occurred are summarized in Tables I and II. All but four analyses were run at 25 percent sensitivity.

---

[1] Technical Equipment Corporation, 917 Acoma Street, Denver, Colorado.

Table I. *Differential thermal analysis of pesticides (temperatures of endothermic and exothermic peaks of reference standards and formulations)*

Pesticide	Sensi-tivity (%)	Temperature (° C.)	
		Endothermic peaks	Exothermic peaks
Phenoxyalkyl acids			
2, 4-D			
Standard	25	165, 310, 479	348, 541
4 lb./gal.	25	154, 300, 363	550
2, 4, 5-T			
Standard	25	168, 309, 504	596, 675
4 lb./gal.	50	687	307, 380, 496, 683
Chlorinated alkyl and aryl acids			
Dicamba			
Standard	25	50, 140, 310, 420, 460	810
4 lb./gal.	25	132, 260, 276, 290, 325	479, 496, 600, 840
Dalapon			
Standard	25	125, 220	
85% W.P.	25	50, 425, 840	210, 382, 471
Tordon			
Standard	25	225, 270, 482	
11.6%			
solution	25	145, 160, 171, 173, 334, 355	190, 386, 475, 580
Chlorinated hydrocarbons			
DDT			
Standard	25	60, 110, 180, 350	545
Technical			
flakes	50	70, 110, 240, 275, 355	375, 578, 825
Dieldrin			
Standard	25	135, 175, 540, 585	325, 360, 420
1.5 lb./gal.	25	50, 240	325, 360, 415, 423, 513
Heterocyclics			
Atrazine			
Standard	25	210, 308, 334, 360, 400, 480	290, 385, 582
80% W.P.	25	182, 280	250, 315, 550
Bromacil			
Standard	25	168, 300, 314, 330	239, 396, 509, 658
80% W. P.	25	170, 250, 307, 370	25, 248, 261, 452, 630
Paraquat			
Standard	25	173, 400, 472, 530	431, 560
2 lb./gal.	25	161, 200, 261, 300, 454	556
Carbamates			
Sevin			
Standard	10	160, 287	656, 683
10% dust	25	159, 169, 249	274, 550
Vernam			
Standard	25	263, 409	
6 lb./gal.	25	61, 196, 369	465

Table I. (Continued)

Pesticide	Sensi- tivity (%)	Temperature (° C.)	
		Endothermic peaks	Exothermic peaks
Phenols			
DNBP			
Standard	25	143, 191, 339, 483	527
3 lb./gal.	25	94, 446	221, 527, 575, 588
Ureas			
Diuron			
Standard	25	173, 330, 400, 492	535, 757
80% W.P.	25	50, 154, 276	293, 460
Organophosphates			
Malathion			
Standard	25	500	250, 308, 333, 422
5 lb./gal.	25	145, 441, 475	261, 308
Miscellaneous			
DSMA			
Standard	10	121, 171, 202, 367, 600, 659	446, 558
3.2 lb./gal.	25	145, 556	388, 470
Nemagon			
Standard	25	50, 196, 358, 396, 432, 456, 500	579, 781
8.6 lb./gal.	25	170, 181	356, 377, 381, 532
PMA			
Standard	25	170, 242, 250, 253	275, 323, 358, 515
95% water dispersable	25	157, 232	259, 311, 334, 376, 400, 563
Treflan			
Standard	25	50, 261, 494, 619	272, 850
4 lb./gal.	25	138, 239	504, 515, 615, 808
Zineb			
Standard	25	185, 496	196, 245, 525, 675, 690, 749, 767, 795
75% W.P.	25	187, 290, 390, 452, 550	215, 232, 620

The 2,4,5-T and DDT formulations and carbaryl and DSMA standards were run at different sensitivities to bring out the peaks more sharply and thereby facilitate the determination of the exact temperature for each peak.

It may be noted from Table I that most of the peaks occurred between 50° and 600° C.

Table II summarizes the temperatures of complete combustion of pesticide materials estimated from the differential thermal curves. The temperature ranges for complete combustion of reference standards

Table II. *Differential thermal analysis of pesticides (temperatures of complete combustion of reference standards and formulations)*

Pesticide	Sensitivity (%)	Temperature (° C.) of reference standard	Sensitivity (%)	Temperature (° C.) of formulation
Phenoxyalkyl acids				
2,4-D	25	602	25	623
2,4,5-T	25	717	50	731
Chlorinated alkyl and aryl acids				
Dicamba	25	840	25	850
Dalapon	25	250	25	850
Tordon	25	550	25	640
Chlorinated hydrocarbons				
DDT	25	560	50	850
Dieldrin	25	620	25	640
Heterocyclics				
Atrazine	25	650	25	600
Bromacil	25	716	25	671
Paraquat	25	613	25	592
Carbamates				
Sevin	10	724	25	678
Vernam	25	447	25	508
Phenols				
DNBP	25	639	25	656
Ureas				
Diuron	25	775	25	550
Organophosphates				
Malathion	25	663	25	715
Miscellaneous				
DSMA	25	665	25	612
Nemagon	25	800	25	596
PMA	25	545	25	646
Treflan	25	879	25	842
Zineb	25	840	25	690

and formulations are from 250° to 879° C. and from 508° to 852° C., respectively. A conspicuous variation in temperatures is particularly evident between the reference standard and its formulation. Dalapon may be singled out as an extreme in this respect. Combustion of the dalapon standard is completed at or about 250° C.; however, the complete incineration of the formulation requires more than a three-fold elevation of this temperature.

It appears from these analyses that temperatures for complete combustion of pesticide materials under study are well within the ranges readily obtainable under most practical conditions. With a very few exceptions, it seems safe to assume that temperatures at or near 1000° C. will be sufficient to degrade 99 percent or more of most commercial pesticidal formulations.

b) Dry combustion

In another research area involving thermal analysis of pesticides, a dry combustion furnace was utilized. The method employed was a slightly modified version of that used for the determination of total carbon by dry combustion (Allison et al. 1965). It involved the burning of 50 to 100 mg. of a pesticide at 900° C. for 15 minutes in a resistance-type furnace in a stream of purified oxygen and the absorption of the carbon dioxide in the effluent gas stream by ascarite. The evolved carbon dioxide was determined gravimetrically. A plug of platinized asbestos serving as a catalyst was inserted in the combustion tube at the rear of the heated zone to insure complete oxidation of carbon monoxide to carbon dioxide. The boat was heated by radiation, conduction and convection in a tube surrounded by heating elements made of high-resistance material. Gaseous products of thermal breakdown of organic molecules other than carbon dioxide which would also be absorbed and could introduce a large error in estimating the carbon dioxide were removed from the air stream. The purification system used to free the incoming and outgoing stream of oxygen was as follows: (1) concentrated sulfuric acid was used to trap ammonia, hydrocarbons, and water vapor, (2) activated manganese dioxide to absorb nitrogen and sulfur oxides and halogens, (3) magnesium perchlorate to pick up water vapor, and (4) ascarite to absorb carbon dioxide.

Table III presents a partial list of total carbon recoveries of analytical grade pesticides. The results show a wide range of recoveries. Roughly ten percent of the carbon was not accounted for as carbon dioxide. Dicamba, atrazine, paraquat, bromacil, and PMA stood out with extremely low recoveries ranging from 67 to 88 percent. These data indicate that under similar conditions of incineration the pesticides may yield volatile carbon products other than carbon dioxide.

c) Ashing (muffle furnace)

Thermal treatment of pesticides was also carried out utilizing a regular combustion (muffle) furnace. Since the commercial formula-

Table III. *Total carbon of analytical grade pesticides*

Pesticide	Carbon (%)		
	Calculated	Determined	Accounted for
Phenoxyalkyl acids			
2,4-D	43.4	41.6	95.9
2,4,5-T	37.6	37.4	99.5
Chlorinated alkyl and aryl acids			
Dicamba	43.2	37.9	87.7
Tordon	29.5	26.9	91.2
Chlorinated hydrocarbons			
DDT	46.8	47.5	101.5
Dieldrin	37.4	36.5	97.6
Heterocyclics			
Atrazine	44.5	31.8	71.5
Paraquat	56.0	37.4	66.8
Bromacil	41.4	35.4	85.5
Carbamates			
Sevin	71.5	65.4	91.5
Phenols			
DNBP	49.5	47.4	95.8
Ureas			
Diuron	46.3	35.5	76.7
Miscellaneous			
PMA	28.2	23.1	81.9
Zineb	17.4	19.4	111.5

tions of pesticides contain certain inorganic materials as carriers or fillers it was desirable to ascertain the quantity of the pesticide mixtures which would be completely combustible at selected temperatures.

The quantity of pesticides lost to combustion at five selected temperatures was determined and is presented in Table IV. It may be noted that the majority of the formulations approach complete combustion at 800° C. However, atrazine, malathion, zineb, bromacil, dalapon, diuron, DSMA, and Sevin have considerable quantities remaining even after combustion at 1000° C. Approximately one-fourth of the zineb formulation is made up of uncombustible materials at the highest temperature tested.

Table IV. *Percent loss on combustion of commercial formulations of pesticides at five temperatures*

Commercial formulation	Loss (%) at				
	600° C.	700° C.	800° C.	900° C.	1000° C.
Picloram	90.8	91.8	95.6	98.7	99.2
Atrazine	87.8	88.1	88.8	88.9	89.0
Nemagon	99.6	99.6	99.6	99.6	99.6
Trifluralin	99.7	99.8	99.8	99.8	99.8
Malathion	95.3	96.0	96.3	96.4	96.7
2,4,5-T	99.9	99.9	99.9	99.9	99.9
Zineb	70.1	71.3	71.5	72.7	72.8
Vernam	99.6	99.6	99.6	99.6	99.6
Paraquat	98.3	98.6	99.0	100.0	100.0
Dicamba	98.6	98.7	98.9	99.0	99.4
Bromacil	88.8	89.1	89.4	90.5	91.3
Dieldrin	99.1	99.4	99.5	99.5	99.5
DDT	99.2	99.3	99.7	99.9	100.0
Dalapon	64.3	64.3	67.8	73.8	91.0
2,4-D	99.8	99.9	99.9	99.9	99.9
Diuron	94.6	95.0	95.4	95.5	95.7
DNBP	99.8	99.8	99.8	99.8	99.8
DSMA	80.6	80.7	80.7	81.2	81.2
Sevin	88.7	88.8	88.8	89.1	89.5

Another phase of the investigations involved exposure of analytical grade pesticides to "thermal shocks"[2] at selected temperatures.

The procedure was as follows: A small sample (0.5 to 1.0 g.) of a pesticide was weighed accurately in a Coors No. 0 crucible and subsequently heated for 30 minutes in a muffle furnace at a desired temperature. Temperatures were selected on the basis of the DTA curves. Following the heating the residue was cooled to room temperature and a small sub-sample removed for infrared analysis. Weights were obtained and recorded after the heating and removal of a portion of the residue for the infrared determinations. The remaining residue was heated to the next higher temperature plateau starting out with a cold muffle and the above-described procedure repeated.

The following pesticides were used in this study: DNBP, dicamba, paraquat, 2,4,5-T, dieldrin, nemagon, zineb, dalapon, malathion, and vernolate. The percentage losses at selected temperatures are summarized in Table V. Under these conditions of heating roughly two-thirds of the weight of most pesticides was lost between 200° and 300° C. In some cases practically all of the weight lost by a pesticide due to heating was in a single 100° C. rise in temperature increment; typical examples are shown in Table V.

---

[2] Heating with intermittent cooling.

Table V. *Effect of heating on weight loss, color, and physical appearance of pesticides* [a]

Pesticide	Heated (° C.)	Loss (%)	Color	Physical appearance
DNBP	200	48.7	Yellowish-brown	Liquid
	250	98.7	Dark-brown	Liquid solidifies on cooling
	300	98.9	Dark-brown	Solid
	400	98.9	Black	Flakes
	500	99.9	Grey	Powder
Banvel D (dicamba)	200	77.1	Colorless	Liquid solidifies on cooling
	250	99.1	None	No residue
	300	99.7	None	No residue
	400	99.8	None	No residue
Paraquat	200	2.0	White	Crystals
	250	2.1	Light brown	Crystals
	300	94.9	Black	Crystals
	400	98.5	Black	Crystals
2,4,5-T	200	27.4	Light brown	Liquid solidifies on cooling
	250	63.6	Light brown	Liquid solidifies on cooling
	300	99.7	Grey	Powder
	400	99.6	Reddish-brown	Powder
	500	100.0	None	No residue
Dieldrin	200	40.8	Yellow	Liquid solidifies on cooling
	250	80.7	Dark-brown	Liquid solidifies on cooling
	300	91.6	Dark-brown	Solid
	350	95.9	Black	Flakes
	400	96.8	Black	Flakes
	500	99.9	Black	Grains
Nemagon	100	71.7	Colorless	Liquid
	150	100.0	None	No residue
Zineb	200	10.0	Brownish-black	Granules
	250	11.9	Brownish-black	Granules
	300	44.8	Black	Granules with metallic sheen
	400	50.3	Yellowish-brown	Granules with metallic sheen
	500	51.9	Yellowish-brown	Granules with metallic sheen
	600	57.6	Yellowish-brown	Granules with metallic sheen
Dalapon	200	100.0	None	No residue
Malathion	200	58.7	Dark-brown	Liquid
	250	72.9	Dark-brown	Liquid, jelly-like on cooling
	300	76.2	Dark-brown	Silk flakes
	350	80.3	Dark-brown	Silk flakes
	400	80.3	Dark-brown	Silk flakes
	500	80.3	Dark-brown	Silk flakes
	600	80.3	Dark-brown	Silk flakes
Vernolate	200	99.6	Grey	Powder
	250	99.6	Grey-brown	Powder
	300	99.9	Brown	Powder
	400	100.0	None	No residue

[a] Analytical standards.

IV. Chemical treatment

Commercial formulations of pesticides were subjected to selected concentrations of five different chemicals with the objective either to partially degrade or to decompose the compounds. The compounds selected were atrazine, malathion, nemagon, carbaryl, dieldrin, trifluralin, bromacil, picloram, diuron, dicamba, DNBP and 2,4-D.

Solutions of the pesticide formulations were treated with three different concentrations of hydrogen peroxide, nitric acid, sulfuric acid, sodium hydroxide, or ammonium hydroxide. Thin-layer chromatography (WALKER and BEROZA 1963) and infrared analysis (SADTLER INDEX 1966) were used to determine the effect of treatment on the pesticides.

The procedure for analysis was as follows: The formulations were either reacted directly or made up in benzene at the rate of 10 mg. of active ingredient/ml. of solution. Chemical agents were prepared as aqueous solutions at strengths shown in Table VI.

Table VI. *Degradation or decomposition reagents utilized*

Reagent	Concentration of solution		
	A	B	C
Hydrogen peroxide	5%	15%	30%
Nitric acid	4N	8N	16N
Sulfuric acid	9N	18N	36N
Sodium hydroxide	2N	4N	8N
Ammonium hydroxide	5N	7.5N	15N

For the benzene solutions an excess amount of the chemical agent to the pesticide was arbitrarily chosen in a 5:1 (v/v) ratio. After addition of the chemical, the mixtures were thoroughly shaken, allowed to stand for five to ten minutes, shaken again vigorously, and allowed to react at room temperature for 24 to 36 hours. The solvent layer was drawn off, evaporated to approximately 1.0 ml. in a Kuderna-Danish evaporative concentrator and analyzed by either thin-layer chromatography or by infrared analysis. The aqueous layer was extracted three times with either chloroform-benzene or n-hexane and subjected to infrared analysis.

The effects of chemicals on pesticides as revealed by infrared analysis may be summarized as follows:

1. Hydrogen peroxide treatment had no significant effect.
2. Nitric acid induced changes in carbaryl and atrazine. The 16N acid treatment produced nitrobenzene from carbaryl and a 2-hydroxy compound from atrazine which is believed to have resulted from the acid hydrolysis of 2-chloro-s-triazine (ZWEIG 1964).

3. Sulfuric acid effected a definite change in the structure of bromacil.

4. Sodium hydroxide caused only slight changes in the structure of dieldrin while malathion was broken down sufficiently to yield inorganic phosphate. Dicamba exhibited no change when treated with sodium hydroxide, yet 2,4-D was hydrolyzed to the sodium salt. DNBP was altered structurally by sodium hydroxide but the spectrum produced is still unresolved. Upon treatment with sodium hydroxide picloram was decarboxylated and the chlorine present was replaced by an OH group. Structural changes were also produced in bromacil when treated with sodium hydroxide.

5. Ammonium hydroxide treatment of carbaryl produced 1-naphthol but effected only a color change in trifluralin.

Summary

Investigations concerning chemical and thermal decontamination of 20 pesticide chemicals are described. These are 2,4-D, picloram, atrazine, diuron, trifluralin, bromacil, DSMA, DNBP, dicamba, dalapon, paraquat, vernolate, 2,4,5-T, carbaryl, DDT, dieldrin, malathion, PMA, zineb, and nemagon.

Differential thermal analysis showed that complete incineration temperatures of the reagent-grade pesticides ranged from about 250° C. to approximately 850° C.; 15 of the compounds were completely combustible at 700° C. or below, while five required 700° and 900° C. Under similar conditions the commercial formulations required essentially the same temperature ranges; dalapon, trifluralin, and nemagon required higher temperatures than the respective reagent-grade compounds.

The incineration of commercial formulations showed a similar temperature range. All but six formulations approached complete combustion at 800° C. Atrazine, carbaryl, bromacil, and dalapon contained about ten percent of uncombustible residue at 1,000° C., whereas DSMA and zineb yielded 19 and 23 percent ash, respectively, at 1,000° C.

Ten compounds, subjected to stepwise heat treatment, showed a temperature range similar to that for differential thermal analysis. Infrared spectra indicated that nemagon disintegrated completely between 0° and 150° C.; whereas dalapon and vernolate did so between 0° and 200° C. Dieldrin, banvel D, and malathion showed no IR absorption after heating to 200° to 250° C. Malathion left a residue of phosphate which remained stable above 600° C. Disappearance of 2,4,5-T and DNBP occurred between 250° and 300° C.; however, DNBP degraded at 250° C. to 2-sec-butyl phenol. Paraquat dichloride degraded between 300° and 400° C. and zineb between 500° and 600° C.; zineb yielded between 200° to 250° C. a thermal degradation product which was tentatively identified as $S = C = N - CH_2 - CH_2 - N = C = S$ or $CH_3 - CH_2 - N = C = S$.

Results from the incineration of reagent-grade pesticides by the dry combustion procedure indicate that on the average ten percent of the carbon is not accounted for as CO_2 at 900° C.

Eight of the pesticide compounds when treated with strong acid or strong alkali exhibited partial decomposition. However, acid or alkaline hydrolysis was not complete in any case for the lengths of time studied.

Results indicate that incineration is superior to chemical methods for the destruction of waste pesticide chemicals.

Résumé *

Méthodes chimiques et thermiques de destruction des pesticides

On décrit des recherches concernant la destruction chimique et thermique des 20 pesticides chimiques suivants : 2,4-D, picloram, atrazine, diuron, trifluraline, bromacile, DSMA, DNBP, dicamba, dalapon, paraquat, vernolate, 2,4,5-T, carbaryl DDT, dieldrine, malathion, PMA, zinèbe et nemagon.

L'analyse thermique différentielle a montré que les températures d'incinération complète des pesticides chimiquement purs se rangeaient entre 250° et 850° C. environ ; 15 d'entre eux étaient complètement combustibles à 700° C. et moins, les cinq autres l'étaient entre 700° et 900° C. Dans des conditions similaires, les formulations commerciales ont nécessité des températures du même ordre de grandeur ; cependant, le dalapon, la trifluraline et le nemagon ont requis les températures plus élevées qu'à l'état pur.

L'incinération des formulations commerciales a mis en évidence une gamme de températures similaire à celle de l'analyse thermique différentielle. Exception faite de six formulations, la température de combustion complète était de 800° C. L'atrazine, le carbaryl, le bromacile et le dalapon contenaient environ 10 pourcent de résidu incombustible à 1000° C., tandis que le DSMA et le zinèbe donnaient respectivement 19 et 23 pourcent de cendres à 1000° C.

Dix composés soumis à un chauffage par paliers ont présenté une gamme de températures similaire à celle de l'analyse thermique différentielle. Le spectre infra-rouge a indiqué que le nemagon se désintégrait complètement entre 0° et 150° C.; le dalapon et le vernolate entre 0° et 200° C. La dieldrine, le banvel D et le malathion n'ont présenté aucune absorption IR après chauffage de 200° à 250° C. Le malathion a donné un résidu de phosphate stable à une température supérieure à 600° C. Le 2,4,5-T et le DNBP avaient disparu entre 250° et 300° C. ; cependant, le DNBP s'était décomposé en 2-butylphénol *sec* à 250° C. Le dichlorure de paraquat s'est décomposé entre 300° et 400° C. et le zinèbe entre 500° et 600° C. Le zinèbe a donné entre 200° et 250° C. un produit de dégradation thermique identifié

---

* Traduit par S. Dormal-van den Bruel

expérimentalement comme $S = C = N - CH_2 - N = C = S$ ou $CH_3 - CH_2 - N = C = S$.

Les résultats de l'incinération par combustion sèche des pesticides chimiquement purs indiquent qu'en moyenne dix pourcent du carbone ne doivent pas être considérés comme du CO_2 à 900° C.

Huit des pesticides ont présenté une décomposition partielle par traitement à l'aide d'un acide fort ou d'une base forte. Cependant, l'hydrolyse acide ou alcaline n'était complète dans aucun cas pour les durées étudiées.

Les résultats indiquent que l'incinération est supérieure aux méthodes chimiques pour la destruction de déchets de pesticides chimiques.

Zusammenfassung *
Chemische und thermische Methoden für die Beseitigung von Pestiziden

Untersuchungen über die chemische und thermische Dekontamination von 20 Pestizidchemikalien werden beschrieben. Diese sind: 2,4-D, "Picloram," Atrazin, Diuron, "Trifluralin," Bromacil, "DSMA," DNBP, "Dicamba," Dalapon, Paraquat, Vernolat, 2,4,5-T, Carbaryl, DDT, Dieldrin, Malathion, "PMA," Zineb und Nemagon.

Differential-Thermalanalysen zeigten, dass Temperaturen für die vollständige Einäscherung von reinsten Pestiziden von etwa 250° bis ungefähr 850° C. rangierten, 15 der Verbindungen konnten vollständig verbrannt werden bei 700° oder niedriger, während fünf 700° bis 900° C. erforderten. Unter ähnlichen Bedingungen erforderten die kommerziellen Formulierungen grundsätzlich dieselben Temperaturbereiche; Dalapon, Trifluralin, und Nemagon erforderten höhere Temperaturen als die entsprechenden Reinsubstanzen.

Die Einäscherung von kommerziellen Formulierungen zeigten einen ähnlichen Temperaturbereich. Alle ausser 6 Formulierungen erreichten vollständige Verbrennung bei 800° C. Atrazin, Carbaryl, Bromacil und Dalapon enthielten etwa 10 prozent von unverbrennbarem Rückstand bei 1000° C., während "DSMA" und Zineb bei 1000° C. 19, bezw. 23 prozent Asche ergaben.

Zehn Verbindungen, welche stufenweiser Hitzebehandlung ausgesetzt waren, zeigten einen ähnlichen Temperaturbereich wie für die Differential-Thermalanalyse. Infrarotspektren deuten an, dass Nemagon sich vollständig zwischen 0 und 150° C. zersetzt, während Dalapon und Vernolat sich zwischen 0 und 200° C. zersetzten. Dieldrin, "Banvel D" und Malathion zeigten keine Infrarotabsorption nach Erhitzung auf 200 − 250° C. Malathion hinterliess einen Phosphatrückstand, welcher über 600° C. hinaus stabil blieb. Das Verschwinden von

2,4,5-T und DNBP wurde zwischen 250 und 300° C. erreicht, jedoch DNBP zerfiel bei 250° C. zu 2-sek.-butylphenol. Paraquatdichlorid zerfiel zwischen 300 und 400° C. und Zineb zwischen 500 und 600° C., Zineb ergab zwischen 200 und 250° C. ein thermisches Abbauprodukt, welches versuchsweise als $S = C = N - CH_2 - CH_2 - N = C = S$ oder $CH_3 - CH_2 - NCS$ identifiziert wurde. Ergebnisse nach der Veräscherung von reinsten Pestiziden mit der trockenen Verbrennungsmethode zeigten, dass im Mittel 10 prozent des Kohlenstoffs nicht als CO_2 nachgewiesen wurde bei 900° C.

Acht der Pestizidverbindungen zeigten teilweise Zerstörung, nachdem sie mit starker Säure oder starkem Alkali behandelt worden waren. Jedoch war die saure oder alkalische Hydrolyse in keinem Fall vollständig für die untersuchten Zeiten.

Die Ergebnisse zeigen, dass Veraschung den chemischen Methoden der Zerstörung von Abfall-Pestizidchemikalien überlegen ist.

References

Allison L. E., W. B. Bollen, and C. D. Moodie: Total carbon. In: Methods of soil analysis, part 2. Agronomy Monograph No. 9. Amer. Soc. Agronomy, Madison, Wis. (1965).

Chesters, G., O. N. Allen, and O. J. Atloe: Differential thermograms of selected organic acids and derivatives. Proc. Soil Sci. Soc. Amer. 23, 454 (1959).

New York State College of Agriculture: A guide to safe pest control around the home. Cornell Misc. Bull. No. 74.

Sadtler Index: Sadtler Research Laboratories, Inc., Philadelphia, Pa. (1966).

Smothers, W. J., and Yao Chiang: Handbook of differential thermal analysis. New York: Chemical Publishing Co. (1966).

U.S. Department of Agriculture, Agricultural Research Service: Safe disposal of empty pesticide containers and surplus pesticides (recommendations). Aug. (1964).

—— Safe use of agricultural and household pesticides. Agr. Handbook No. 321.

Walker, K. C., and M. Beroza: Thin-layer chromatography for insecticide analysis. J. Assoc. Official Agr. Chemists 46, 250 (1963).

Zweig, G. (Ed.): Analytical methods for pesticides, plant growth regulators and food additions, Vol. IV. New York: Academic Press (1964).

Some research approaches toward minimizing herbicidal residues in the environment

By

C. L. Foy * AND S. W. BINGHAM *

Contents

I. Introduction

According to a recent quotation, "starvation and environmental deterioration are the inevitable results of a burgeoning world population. To consider this fact only 'a problem of the underdeveloped world' is like saying to a fellow passenger 'your end of the boat is sinking.'"

Paradoxically, in some instances, the very tools of advanced technology that go hand in hand with increased agricultural production, and that effectively combat starvation, may themselves actually become sources of environmental contamination.

The economic value and world-wide importance of chemical pesticides as agricultural production tools are unquestioned. However, residues of pesticides and other exogenously applied chemicals in foodstuffs are of concern to everyone everywhere; they are essential to food production and manufacture, yet without knowledge of their

---

* Department of Plant Pathology and Physiology, Virginia Polytechnic Institute, Blacksburg, Virginia.

behavior in the environment, proper surveillance, and intelligent control, certain of the persistent pesticides could at times conceivably endanger the public health.

We are all consumers of air, water, and food. Since we also occupy and partake of the earth's environment, it behooves every person to develop an awareness and appreciation of the hazardous potential created by chemical contamination of the environment by any means. The truism that "no man is an island" is thus brought into even sharper focus as the world population continues its rapid increase. Also, man alone, the manipulator of nature, must assume responsibility for "environmental sanitation" if it is ever to be practised. In this respect, then, to a very real extent, each man is his "brother's keeper."

Public concern over the potential health hazards of certain pesticides in recent years has given impetus to research aimed at preventing, minimizing, or removing such chemical residues from the biosphere. Conceivably, all segments of the biosphere — air, soil, water, and ultimately the plants and animals that occupy space therein — may be subject to chemical contamination through promiscuous and unwise use of chemical pesticides. Thus, it is necessary to establish acceptable chemical tolerances in specific agricultural (food and feed) commodities as listed in the current U.S. *Department of Agriculture* "Summary of Registered Agricultural Pesticide Chemical Uses" (*Pesticide Regulation Division* 1968); it is also important that we take an overall view of the problem of pesticide distribution and re-distribution in nature.

In general, herbicides as a use group are less hazardous to man and animals than are many of the insecticides, for example. Some are virtually innocuous, *i.e.*, "nontoxic" at any practical level; however, there are enough important exceptions that are rated as "moderately toxic" and "very toxic" to warrant caution and concern in the use of all herbicides. Indeed this should be true in the use of all pesticides and other foreign chemicals, in view of the fact that their long-term effects are often incompletely understood.

In this paper, we will first discuss, briefly, the general distribution and fate of herbicides in the environment following their application to air, plants, soils, and water. Next we will consider the normal modes of dissipation of herbicides, particularly from soils. Finally, several possible approaches toward preventing, minimizing, or removing herbicide residues from the environment will be suggested and discussed, insofar as available data will permit. Hopefully, the suggestions and examples cited will serve to promote further research in the area of environmental sanitation, both preventive and corrective.

II. Scope, distribution, and fate of herbicides

Weed Science is more than just the destruction of undesired

vegetation — or plants out of place. It is a rapidly developing new scientific discipline involving the study of weeds and their control by cultural, mechanical, biological, chemical, and combination methods in field crops, horticultural crops, pastures and rangelands, forests, and non-cropland areas. In all control methods involving the application of chemical herbicides and spray additives, the possible occurrence of chemical residues in the environment becomes a public concern.

Thus weeds and their control through the use of chemicals concern not only the large land owner, the small farmer, and the owner of urban lawns and gardens, but also the city dweller who merely purchases his food from the supermarket. Fruit growers, turf specialists, golf course superintendents, livestock ranchers, irrigation specialists, rights-of-way maintenance personnel, foresters, maintenance engineers, farm implement manufacturers, seedsmen, warehousemen, synthetic organic chemists, analytical chemists, public health officials, military officials, and housewives also feel the impact of weeds and herbicide uses upon their lives. Truly, these matters are everybody's business.

Contrary to some indications, both given orally and seen in the press, research scientists — representative of the agricultural chemicals industry and other users of pesticides — are no more anxious to denude the land and poison themselves out of existence than are any other groups selected from the consumer population.

The use of pesticides in the United States continues to expand rapidly. Herbicides and plant hormones accounted for 54.6 percent of the 1967 total dollar value at the primary producer's level, reflecting a 271 percent increase in the sale of herbicides since 1963 (MAHAN *et al.* 1968). The Weed Science Society of America now lists approximately 120 organic herbicidal compounds. Herbicide sales will undoubtedly continue at a high rate, taking over an even greater share of the pesticide market. Thus research attention on matters pertaining to herbicide residues is amply justified.

a) General distribution and fate in the environment

The fate of herbicides in plants, soils, air, and water is a relatively recent subject for scientific investigation. The accumulative evidence today indicates that only a relatively minute portion of the chemical that is applied may actually arrive at the site of action and function herbicidally (Fig. 1). Morphological, soil, translocation, physiological, and biochemical "obstacles," and their interactions with environmental influences, determine selective toxicity by affecting the concentration of the herbicide that reaches the site of action (SHAW *et al.* 1960). These factors also influence, to a marked degree, the fate of the herbicide applied that is extraneous.

Figure 2 depicts schematically the possible distribution and fate of herbicides applied to air, plants, soil, and water. Such organic chem-

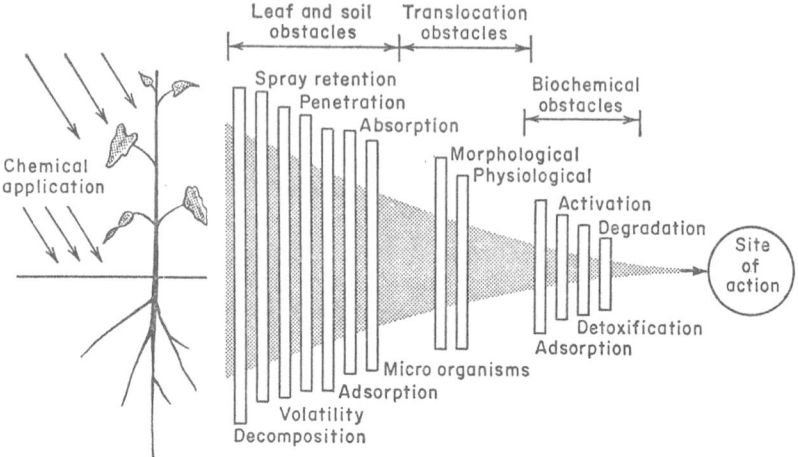

Fig. 1. Obstacles which determine the concentration of toxic material at the site of action (Shaw et al. 1960)

icals introduced into a plant, or into soil or water are, of course, subject to possible degradation, alteration of structure, or conjugation with natural constituents. Thus flow lines on the scheme presented (Fig. 2) may depict the fate of metabolites or decomposition products of herbicides instead of, or in addition to, that of the original molecules. Such compounds, *i.e.*, products of herbicide degradation or conjugation with natural constituents, are not distinguished for purposes of this discussion. However, as Schechter (1968) has pointed out "With the increased amount of analytical work being done in the surveillance and monitoring of pesticides in our environment and on residues in general, it seems appropriate to reiterate the importance of confirming the identity of pesticides being reported. Confirmation is particularly important with samples from the environment when little or no information is available concerning the pesticides likely to be present."

In certain instances, herbicide activation rather than detoxification occurs in plant tissues and/or soils. Some examples are 2,4-DB,[1] 2,4-DEB, 2,4-DEP, 2,4,5-TES, MCPB, and MCPES, which are either known or believed to be converted to their corresponding herbicidal acetic acid derivatives 2,4-D, 2,4,5-T, and MCPA.

It is apparent from Figure 2 that general fumigants such as methyl bromide and chloropicrin, that are usually administered under air-tight cover, enter the atmosphere directly as gases. Sprayed herbicides and even granular and dust materials applied from some height pass through the air and, thus contribute "fines," micronized particles,

---

[1] Herbicides and other compounds mentioned in the text are identified chemically in Table VII.

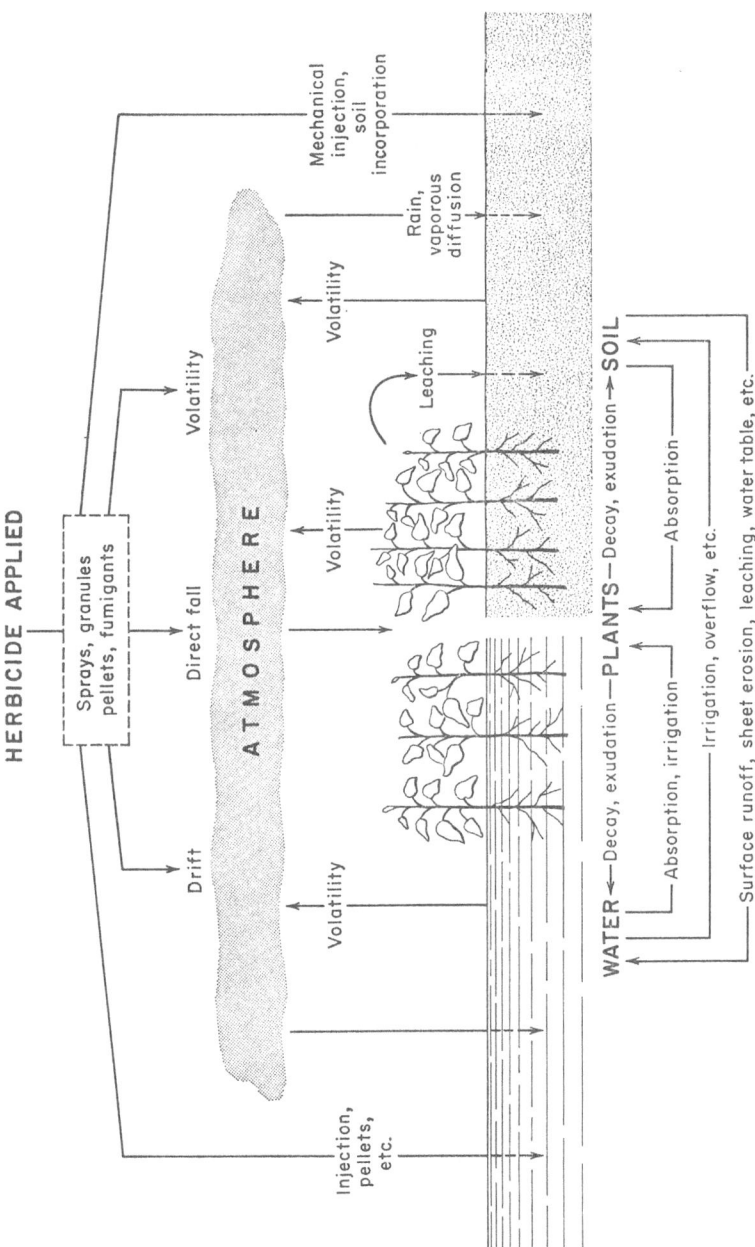

Fig. 2. Scheme showing the possible distribution and fate of herbicides and their degradation products in the biosphere

and/or "volatiles" to the atmosphere. After a herbicide has already contacted the soil, plants or other objects, volatile chemicals (either the starting herbicide or its degradation products) may re-enter the atmosphere in vapor form.

Some herbicides such as acrolein, endothall, paraquat, and simazine are administered directly into water for aquatic weed control. Numerous other herbicides enter creeks, streams, rivers, and standing bodies of water by atmospheric fallout, sheet erosion of soil, and surface runoff and leaching from plants and soils. Although not studied in any detail yet, certain highly mobile and persistent herbicides such as dicamba and picloram might conceivably be leached into the subsoil, enter the water table, and thus be widely re-distributed. This possibility is considered highly unlikely, however, in view of the fact that all organic herbicides known thus far are eventually dissipated in nature, most of them fairly rapidly. Compared with DDT, a so-called "hard" or persistent pesticide, most herbicides are relatively "soft" or non-persistent in living organisms or in nature. Even high application dosages of the "soil sterilant" chemicals (e.g., monuron, diuron, bromacil, simazine, and atrazine) eventually disappear usually after two to three years. Most commonly, the longevity of organic chemicals applied at herbicidal levels to the soil or water is measured in months, weeks, or sometimes, even days.

Certain herbicides and growth regulating chemicals, once absorbed by plant shoots and/or roots, are degraded rapidly creating no residue problem; others are remarkably mobile and persistent. Thus plants are capable of releasing these compounds, in nonmetabolized form, back into the environment by exudation and/or leaching. Some exogenous plant growth-regulating compounds that are exuded or excreted from plant roots following application to the foliage are MOPA, dicamba, 2,3,6-TBA, 2,3,5,6-TBA, PBA, and picloram (Preston et al. 1954, Linder et al. 1958 and 1964, Mitchell et al. 1959, Hurtt and Foy 1965 a and b). Other examples could be cited. Herbicide residues trapped within plant tissue, or metabolites thereof, may of course enter or re-enter the soil, water or atmosphere upon death and decay of the plants.

b) Modes of dissipation from soils

Because of the relative residual lives of most herbicides and where and how they are used, the soil is perhaps the largest most important temporary reservoir for the accumulation of herbicide residues. Some portions of practically all herbicides used, whether soil-active or foliarly-applied compounds, eventually contact the soil. Herbicides reaching the soil accidentally or by design, normally become dissipated or removed with time, in one or more of several ways: (1) volatilization, (2) photo-decomposition, (3) soil adsorption-inactivation, (4) leaching, (5) chemical breakdown, (6) microbial degradation, (7) plant uptake, followed by metabolic degradation and/or physical removal at harvest, and (8) sheet erosion-surface removal.

In most critical studies conducted thus far on herbicide disappearance from within soils, degradation by soil-borne microorganisms has proven to be very significant — perhaps indeed the most important single factor in many instances. Conceivably, however, each of the aforementioned modes of dissipation might be subject to deliberate manipulation with a view toward minimizing residue accumulation as well as altering herbicidal action and selectivity. Several such approaches were mentioned recently by LANGE (1967); others are suggested from the literature and from practical experience.

III. Current research approaches toward minimizing herbicidal residues

Basically, there are three main approaches toward minimizing residue accumulation of pesticide chemicals: (a) use a non-chemical substitute method, (b) enhance the toxicity, selectivity, or effectiveness of chemical means so that less pesticide is actually required, and (c) effect the disappearance of the chemical residue, by deliberate means, once contamination has occurred. The first two alternatives are preventive, the last corrective; each approach has merit and is worthy of consideration.

a) Use of alternative methods of vegetation control

It is somewhat trite and obvious to say that the best way to minimize herbicide residues is to refrain from using herbicides. However, several non-chemical methods of vegetation control do have proven merit (CRAFTS and ROBBINS 1962, KLINGMAN 1961) and may well be used to alleviate chemical residue problems in some specific situations. Detailed treatment of such non-chemical approaches to vegetation control is beyond the scope of this report. For purposes of the present discussion, it is sufficient simply to list and recognize several familiar possibilities: (1) cultural methods, including cropping and competition, water management and fertilization practices; (2) mechanical methods including hand pulling, hoeing, tillage, mowing or cutting, flooding, smothering (non-living mulch materials), oiling, and burning (flame cultivation); and (3) biological methods, involving the use of parasitic insects, goats, geese, etc.

b) Minimizing residues by increased effectiveness and selectivity

The literature dealing with efforts toward improving herbicidal effectiveness and selectivity is very extensive. Comprehensive review of all facets is beyond the scope of this paper. However, it is relevant to mention that many technological and physiological factors relating to the improved efficiency of herbicide uses also, coincidentally, bear directly on the chemical residue problem. Frequently, when herbicide effectiveness or its efficiency of use in the field is improved, less of the chemical is actually required, and the residue potential is thereby reduced.

As in the discussion on alternative (non-chemical) methods, numerous examples could be cited whereby, conceivably, chemical residue levels could be reduced. However, little of the published work in this area has been conducted with alleviation of residue problems as the principal consideration in mind.

In this brief report, only selected examples most familiar to the authors will be presented. For a more comprehensive treatment of herbicide physiology and weed control technology, the reader is invited to consult the several recent texts, reviews, and other publications that are available on this broad subject (e.g., Woodford et al. 1958, Currier and Dybing 1959, Woodford and Sagar 1960, Ashton et al. 1961, Klingman 1961, Crafts and Foy 1962, Crafts and Robbins 1962, Ebeling 1963, Hilton et al. 1963, Audus 1964, Foy 1964, Foy et al. 1967, Moreland 1967, Foy and Smith 1968, Kearney and Kaufman 1969).

1. Optimum placement of herbicides. — The principal factors that are recognized to contribute toward herbicidal selectivity are: (a) leaf properties, (b) location of the growing points, (c) growth habits, (d) absorption, (e) translocation, (f) biophysical-biochemical, (g) position of the herbicide in the soil, and (h) selective placement (Ashton et al. 1961). Herbicides can be applied either to plant foliage or to the soil or water in such a manner as to optimize effectiveness and/or selectivity. Such selective placement can be accomplished by the use of banded sprays, shielded sprays, directed sprays, granular or pelleted formulations, specific placement in the soil profile by injection or mechanical incorporation, or other methods. In any case, if less herbicide can be used in effecting the desired degree of vegetation control, it follows logically that the hazard of chemical residues will also be diminished.

2. Reduction of herbicide loss by volatilization. — Volatility may be considered from two viewpoints with respect to possible residue problems. In the first instance, volatility may be considered a wasteful property; i.e., because the herbicide is so rapidly lost, presumably a higher dosage than of a non-volatile formulation of equal potency would be required for sustained control. An additional problem is that the escaped herbicide in the vapor state is then present in the atmosphere and may create residue problems, as well as possibly causing injury to desirable plants at some point remote to the area of application. Volatile forms of the phenoxy herbicides have been particularly objectionable at times.

The rapid loss in vapor form of certain soil-active herbicides such as EPTC and pebulate, particularly from moist soil, necessitates their being incorporated into the soil soon after application (GRAY and WEIERICH 1965). This serves as an effective practical means of reducing herbicide loss by volatility in the field (GRAY 1965). For some herbicides that are known to undergo photodecomposition (e.g., trifluralin), soil incorporation provides the additional advantage of protecting the herbicide from exposure to sunlight.

An obvious practical solution to volatility loss is the selection of a non-volatile formulation to do the job, whenever practicable (e.g., the acid or amine salt rather than the isopropyl ester of 2,4-D). Another experimental approach that has been suggested to reduce volatility loss of herbicides from plant or soil surfaces is the use of anti-transpirants, filming agents, and other additives. However, these as yet have no demonstrated practical importance.

An opposite view on volatility is that this property may be sought in herbicides and used advantageously under certain circumstances, *i.e.*, where the desired herbicidal effects may be achieved relatively quickly then the unneeded excess is dissipated to the atmosphere. Although the extraneous chemical is not degraded immediately, it leaves the immediate scene and creates no local residue problem. The rationale, apparently, is that "The solution to pollution is dilution!".

3. **Enhancement of herbicide penetration and translocation.** — It has long been recognized that suitable surfactants may facilitate and accentuate the emulsifying, dispersing, spreading, wetting, solubilizing, and/or other surface-modifying properties of herbicidal formulations to bring about enhancement of penetration and herbicidal action [CURRIER and DYBING 1959, JANSEN *et al.* 1961, CRAFTS and FOY 1962, McWHORTER 1963, FOY 1964, JANSEN 1964 a and 1965 a, HILL *et al.* 1965, FOY and SMITH (1965 and In press), and many others].

Similar enhancement of herbicidal action by the use of phytobland oils, oil-emulsifier combinations, special organic solvents (e.g., dimethyl sulfoxide), and other spray additives has also been reported (BANDEEN and VERSTRAETE 1967, COATS and FOY 1967, AYA and RIES 1968, FOY and COATS 1968, JONES and FOY 1968, and others).

Used carefully, proper surfactants and other additives offer the possibility of more effective weed control using lower rates of herbicide and, consequently, with fewer residue problems. By making the herbicides more efficient through better coverage and improved penetration, it should be possible to reduce recommended herbicide rates accordingly. Experimentally, this approach has been demonstrated many times. However, in practice, weed specialists are still somewhat reluctant to recommend lower rates of herbicides until plant-herbicide-surfactant-environment interactions are better understood.

c) Minimizing residues by removal, inactivation, or alteration of persistence

The recognized principal modes of herbicide dissipation have been enumerated [see section II b)]. Deliberate manipulation or control of these factors has been attempted in several instances, usually in an attempt to shorten the persistence time.

1. **Volatility.** — Day (1961) investigated two means of reducing soil contamination resulting from the leaching of spray residues of dalapon from plant foliage: use of volatile forms of the herbicide, residues of which slowly evaporate from the foliage, and use of formulations that congeal to a solid film to retain the herbicide against leaching or volatility. Rates of evaporation from glass slides at 20° C. for the ethyl ester, *n*-butyl ester, acid, and diethylene glycol *bis*-ester of dalapon were respectively 15, 1.6, 0.27, and 0.01 mg./cm.^{2} As foliar herbicides the ethyl and butyl esters were ineffective and the acid of limited effectiveness due to rapid evaporation. The glycol ester was as effective as the sodium salt. With its evaporation rate of approximately one kg./ha./hour at 20° C., application of dalapon in the form of the diethyleneglycol *bis*-ester of dalapon with polyvinylacetate emulsion dried to form films that greatly reduced leaching of the herbicide at the expense of minor reduction in herbicidal activity. Such formulations, according to Day (1961), afford a means of limiting soil contamination.

Several physical-chemical and environmental factors influence the vapor loss of such compounds as IPC, CIPC, and other carbamates and thiocarbamates. For example, when vapor losses of IPC and CIPC were studied under controlled laboratory conditions, IPC was more volatile than CIPC; temperature, air flow rate, moisture, and cation-exchange capacity were important factors influencing volatility (Parochetti and Warren 1966). Vapor loss increased with increasing air-flow rate and temperature; losses from moist soil decreased as the percent clay, organic matter, or both, and the cation-exchange capacity increased.

The most important factor affecting the loss of EPTC from soil was the moisture content of the soil (Gray and Weierich 1965). The amount of EPTC lost from dry soils was much less than from moist soils. In another study (Gray 1965), allowing the surface of freshly worked moist soil to dry out to a depth of ½ inch before spraying EPTC greatly reduced the loss by vaporization. Immediate incorporation prevented the loss of EPTC from dry soil and greatly reduced the loss from moist soils. After spraying EPTC on dry soil, sprinkling with small amounts of water increased the loss. Thus it is apparent why incorporation of many herbicides into the soil is done to prevent vapor loss. On the other hand, in areas treated with a space fumigant under a vapor proof barrier, the soil is sometimes stirred afterward as an aid to dissipation by vapor loss.

2. Photodecomposition. — Many herbicides representing widely diverse chemical groups are susceptible to photodecomposition under appropriate circumstances, e.g., several phenylureas (monuron, diuron, fenuron, neburon), s-triazines (simazine, atrazine, ametryne), substituted uracils (isocil, bromacil), amiben, CDEC, diquat, paraquat, etc. (SHEETS 1963, JORDAN et al. 1964 and 1965, COMES and TIMMONS 1965, TAYLORSON 1966, FUNDERBURK et al. 1966, and others).

The studies reported have involved breakdown both in sunlight and in ultraviolet light. Generally, the far UV is more destructive than near UV. The practical importance of photodecomposition of herbicides under field conditions has not been fully assessed. Conceivably, this principle might be employed to degrade herbicides on the soil surface, relating it to land use and management practices, but no purposeful attempt has yet been made in this direction to our knowledge.

3. Alteration of herbicide adsorption and leachability in soils. — Herbicides vary widely not only in their volatility properties but also with respect to solubility, adsorption, and leachability in soils (DONALDSON and FOY 1965, HARRIS 1966, KOREN et al. 1968, RODGERS 1968, and many others). Recently, HARRIS (1966) compared the mobilities of 28 herbicides (new and old) in soil columns, and established "relative mobility factors" for each. The aromatic acid herbicides were generally most mobile, and the insoluble toluidines least mobile in two different soil types. For example, assigning monuron a value of 1.0, the relative mobility factors for 2,3,6-TBA and dipropalin were 2.2 and 0.5, respectively. This concept of data presentation should prove useful in later research aimed toward comparing, understanding and manipulating the mobility of herbicides in soils, with a view toward leaching out undesirable chemical residues.

Decontamination of herbicide residues by means of controlled irrigation practices alone, or in combination with tillage, cropping and the use of soil amendments has been demonstrated experimentally (LANGE 1967). For example, heavy irrigations have been used to leach herbicides out of the root zone of desirable crops. Repeated tillage several times deeper than normal after cropping is somewhat similar in principal, i.e., involving removal or dilution of the herbicide to innocuous levels in the crop root zone, and is sometimes used in combination with irrigation.

Many recognized factors such as soil type (texture, structure, series), cation-exchange capacity, amount and type of clay, organic matter, pH, temperature, moisture equivalent, exchangeable bases, fertility, amount and distribution of rainfall or irrigation, as well as the properties and amounts of the herbicide, solvent, and other additives, may influence adsorption, mobility, and persistence. Numerous studies also attest to the dominant role of microorganisms in the breakdown and removal of herbicides from soils. It is beyond the scope of this presentation to discuss these factors in detail. However, a few selected publications cited for further reference are UPCHURCH 1958, ALEXANDER and ALEIM 1961, SHEETS et al. 1962, UPCHURCH and MASON 1962, BURNSIDE et al. 1963, DAY et al. 1963 and 1968, HANCE 1965, HARRIS and SHEETS 1965, MCRAE and ALEXANDER 1965, KEARNEY 1966, DARDING and FREEMAN 1968, KEARNEY and KAUFMAN 1969.

Continually moist soils have often given more rapid breakdown of some herbicides. This is undoubtedly due in part to the creation of favorable conditions for microbiological activity. However, it has also been suggested (LANGE 1967) that non-irrigated herbicide-treated soils may be capable of "storing" the herbicide unaltered for long periods until such time as water is applied. The process is envisioned as one of temporary binding of the herbicide by adsorption on internal clay lattices and consequent "trapping" during drying and shrinkage. This suggests that re-wetting would thus be a practical approach toward releasing and removing such residues from soils. Convincing experimental evidence for the operation of this mechanism is still needed.

A relatively new approach that has shown promise for influencing or controlling the mobility of herbicides in soils is the use of surfactants or other additives. Surfactant products are now being investigated as an aid to water penetration (PELISHEK et al. 1962). If proven successful, this approach would have particular applicability in certain western soils that are low in permeability and/or high in salts or alkali.

Relatively little information is available concerning the modifying effects of chemical adjuvants (additives) on the mobility, availability and persistence of herbicides in soils (HOWARTH 1962). Still, such influences almost certainly do exist. It is not unexpected that chemical additives applied to the plant either before, with or after herbicide treatment, can modify adsorption, absorption, mobility, leaching, volatility, diffusion, accumulation, and metabolism of the compound in such a way as to increase or decrease herbicide residues. One obvious effect of a surfactant might be the (a) immobilization (inactivation) of a herbicide within a given stratum of the soil, (b) uniform distribution of the herbicide through the profile, thereby accomplishing a desired dilution effect, or (c) complete removal of the herbicide from the root zone by leaching. The desirability of either approach depends on many circumstances, of course.

There are some indications that the use of surfactants might be feasible for this purpose. For example, BAYER (1967) and SMITH and BAYER (1967) found that the depth of leaching of diuron in Yolo soils was altered by the various types — anionic, cationic, or nonionic — surfactants included in the study. While adsorption of diuron was great enough to resist leaching when applied with the surfactants Aliquat 204 and 221, the diuron was still available for absorption by plant roots. Other surfactants of the Aliquat series containing lauryl, dioctyl, and distearyl substitutions had little or no effect on this sorption phenomenon. Thus, the sorption complex appeared to be closely associated with the dilauryl portion of the surfactant.

Definite structural requirements relating to activity [JANSEN 1964 a and b and 1965 a and b, SMITH and FOY 1966 a and b, FOY and SMITH (In press)], translocation (FOY and SMITH 1966 c), and biodegradability (McCUTCHEON 1964) appear to exist for certain chemical groups of surfactants. Whether these also influence herbicide persistence in soils and plants requires further study. Little is known about the alteration of energy relationships at interfaces within the soil by surfactants and other soil additives. When properly understood these factors could be of considerable importance in regulating the depth of penetration, activity and persistence of herbicides in soils.

4. Herbicide inactivation in soils by use of adsorbents. — Herbicidal chemicals in increasing quantities are being applied to soils to prevent or reduce the growth of weeds. In addition, large quantities of these chemicals from postemergence foliar sprays reach the soil. The action of the herbicides is not always limited to the immediate objective of controlling a particular plant. Sensitive plants may be killed or temporarily effected for varying lengths of time. Certain soil properties have a pronounced effect on herbicidal activity. Phytotoxicity with certain herbicides was related inversely to organic matter content of the soil (UPCHURCH 1958, UPCHURCH and MASON 1962, BINGHAM 1967). This fact suggests that adsorbents may play an important role in protection of plants from herbicides and detoxification of residues in soils.

The greatest amount of effort on adsorbents for herbicides has been devoted to activated charcoal. Soon after 2,4-D was recognized to be an active herbicide, activated charcoal was found to inactivate this chemical in a contaminated sprayer and reduce the carryover to sensitive plants (LUCAS and HAMMER 1947). Then it became evident that certain plants sensitive to 2,4-D were protected from injury by dipping or dusting their roots with activated carbon prior to transplanting into treated or contaminated soil (ARLE et al. 1948). Herbicide selectivity in horticultural crops was improved considerably by using a coat of activated charcoal on the roots of various transplants (AHRENS 1965 a and b). The degree of protection afforded by the carbon root dips varied greatly with the crop. Carbon was most effective for protecting the crops that were normally moderately resistant or only slightly injured by the chemical. Tomato plants were quite sensitive to triazine herbicides and were not protected from severe injury with carbon. Selective weed control was attained only with those crops which have some degree of tolerance beforehand.

The action of activated charcoal is envisioned as simple adsorption. Adsorption occurs when a herbicide becomes attached to the surface of another material. Activated charcoal has an extremely large surface area in comparison to its volume and is a very efficient adsorbent. It is extensively used for purification of drinking water, air, cleaning fluid, sugar, and cigarette smoke. Charcoal is used in many gas masks because of tremendous adsorbent qualities.

During recent years, many types of herbicides have been introduced. Studies involving the activated charcoal dip of strawberry roots in soil containing various amounts of 33 different herbicides has demonstrated that broad spectrum protection is possible (Table I).

Table I. *Tests with varying rates of 33 herbicides on two varieties of strawberries (Midway and Surecrop) show how applying activated charcoal to plant roots helps prevent herbicide injury* (SCHUBERT 1967)

Herbicide rate	Charcoal	No charcoal
	Av. wt. of plants (g.)	
0	19.6	19.3
Normal [a]	24.3	17.7
2X Normal	24.4	15.2
5X Normal	21.9	11.3
	Av. no. of new primary roots	
0	11.6	15.1
Normal	18.0	13.1
2X Normal	18.1	11.8
5X Normal	15.9	7.8
	Rating of sec. root developments [b]	
0	6.6	7.1
Normal	7.5	5.3
2X Normal	7.2	4.5
5X Normal	6.2	3.0

[a] Normal rates of 33 different herbicides.

[b] The secondary root development was rated from zero to 10. A rating of zero indicates no new secondary roots and a rating of 10 indicates a very well-branched secondary root system.

This was indicated by plant weights and development of roots (SCHUBERT 1967). In many cases, with two to five times the normal herbicide application for weed control in crops, the carbon-treated plants often grew as vigorously as plants without herbicide. These results indicate that charcoal would prevent injury to moderately sensitive strawberry plants from herbicides used on a previous crop.

During the last few years, the effect of residues of herbicides used for crabgrass control in turfgrasses has been investigated in relation to reseeding an area relatively soon after the application (BINGHAM and

SCHMIDT 1964, JAGACHITZ and SKOGLEY 1966). These studies demonstrated that turfgrasses cannot be seeded safely for a period of a few weeks to several months later, depending on the herbicide used. The example of the herbicides for crabgrass control in turfgrasses is used here since the level of treatment in this case is sufficiently high to give control for the entire year. Herbicide residues present a particular problem in this instance when it becomes necessary to seed new turfgrass. The results of inactivation studies with bensulide using activated charcoal are presented (Table II).

Table II. *The influence of activated charcoal for inactivation of bensulide in soil*

Bensulide (lb./acre)	Inactivation (dry wt. as g./12.56 in.²) [a]				
	Activated charcoal (lb./acre)				
	0	75	150	300	600
Crabgrass					
0	1.6a	1.9a	1.2a	1.3b	2.8ab
5	1.7a	1.2ab	1.9a	2.5a	2.6ab
10	0 b	0.8bc	1.7a	1.8b	2.1b
20	0 b	0.1c	1.1a	1.3b	3.1a
40	0 b	0 c	0 b	0 c	0.2c
Bluegrass					
0	0.2	0.2	0.2	0.2	0.2
5	0.1	0.3	0.2	0.3	0.2
10	0	0.2	0.2	0.1	0.3
20	0	0	0	0.1	0.2
40	0	0	0	0	0
Kentucky 31 fescue					
0	0.8ab	1.0ab	0.7	0.8	1.0
5	1.2a	0.8abc	1.0	0.5	1.2
10	0.5ab	1.2a	0.9	0.9	1.2
20	0.3b	0.3bc	0.4	1.2	1.2
40	0 b	0 c	0.4	0.4	0.4

[a] Weights in the same column for each grass with the same letter are not significantly different (0.05 percent level).

The normal level of treatment is 10 lb./acre for season-long crabgrass control. Application of activated charcoal to the soil surface at 150 lb./acre at 25 days after herbicide treatment reduced the control of crabgrass with even 20 lb./acre of bensulide. Doubling the level of charcoal did not deactivate twice as much bensulide under field conditions on a sandy loam soil.

Kentucky 31 fescue was more tolerant to bensulide in the beginning; however, a similar response to carbon applications was obtained. Common Kentucky bluegrass with a much slower growth rate was slightly more sensitive than crabgrass to bensulide and also responded to activated charcoal indicating that the herbicide became fixed on its surface. These data demonstrate, in the event that reseeding an area to turfgrasses becomes necessary after bensulide has been applied, that activated charcoal can be used effectively to inactivate the herbicide for seedling establishment.

Until recently the problem of reseeding turfgrasses following herbicide treatments for broadleaved weed control was not too severe. The phenoxy herbicides were degraded or otherwise lost from the soil in a few weeks to sufficient levels to allow reseeding the area. The introduction of more stable herbicides such as dicamba into the turfgrass programs revived this problem. JAGSCHITZ (1968) demonstrated that charcoal may be used effectively at the time of seeding the turfgrass to nullify the harmful effects of broadleaf herbicides including dicamba and picloram. The establishment of all the grasses was inhibited or prevented by crabgrass and broadleaf herbicides and was improved in herbicide treated soil by the use of activated charcoal. Estimates of turfgrass response indicate that the seedling stands were similar in treated and untreated soil provided charcoal was used following the treatment. As yet an unanswered question is related to the length of time required before effective pre-emergence control of crabgrass can again be attained using low levels of herbicides after activated charcoal has been applied to the soil.

Some research has been keyed toward the determination of carbon-to-herbicide ratios necessary to completely detoxify a material in order to grow another crop. AHRENS (1965 a and b) found that the amount of inactivation required varied considerably with the crop involved. For certain triazine herbicides, activated charcoal protected certain crops from noticeable injury at a 200-to-one ratio of carbon to herbicide; however, the more sensitive the crop the higher this ratio must be for good protection. As suggested previously, the organic matter content of the soil influences the level of phytotoxicity of most herbicides and influences directly the amount of charcoal required to inactivate a given dose (ANDERSON 1968).

In certain instances the coating of a seed is enough charcoal to afford protection from injury. Wheat may withstand higher levels of chloroxuron and simazine if the seed is initially coated with activated charcoal (SCHUBERT 1967). As more specific information about microorganisms which degrade organic herbicides becomes available, it may be possible to use both charcoal and microorganisms to protect plant roots for a longer period than by charcoal alone. In many instances the plant was protected until roots grew beyond the charcoal region, but it began to develop injury symptoms after roots entered the unprotected soil.

Another way adsorbents may be used is to band them over the seed at planting time (LINSCOTT and HAGIN 1967). Narrow bands of activated charcoal sprayed on the soil over alfalfa seed prevented the normal injury from one triazine herbicide. Other more injurious triazines caused injury regardless of the amount of charcoal applied.

Adsorbents potentially have great value for modern agriculture: (1) to reduce the waiting period normally necessary for planting a crop after herbicides have been used to control weeds, (2) to allow a moderately tolerant transplant crop to avoid injury during the critical establishment period, (3) to permit higher rates or more effective herbicides to be used for hard to kill weeds before planting specific crops, and (4) to deactivate a herbicide if inadvertently applied to the wrong crop. Herbicides have been effectively removed from water with activated carbon (HYNDSHAW 1962). Many tastes and odors traced to the presence of organic herbicides were effectively reduced by the carbon. It is even visualized that animals fed a continuous minimum level of an adsorbent would not accumulate, become contaminated by, or yield products with pesticide residues.

5. **Use of specific trap plants to remove herbicides from water and soils.** — Increasing numbers of herbicides have been demonstrated to give good control of a wide spectrum of aquatic plants. These herbicides may present a direct residue problem in water intended for later uses in irrigation or as drinking water for man or animals. During treatment of large land areas with herbicides for weed control in many crops, surface water runoff, especially during sudden heavy rains, may pose residue problems in water down stream. The more common situation involved in many cases deals with the herbicide residues remaining for extended periods beyond the time required for weed control. Thus, planting a crop or a succeeding crop has to be delayed for a year or more. There have been several research approaches to reduce the residue problem both in water and soil.

GRZENDA et al. (1966) evaluated the persistence of herbicides in pond water and demonstrated that diquat and paraquat persisted for less than a month. On the other hand, fenac and amitrole were detected for more than six months after treatment. FRANKS et al. (1967) studied the water residue problem in water flowing over herbicide-treated irrigation canals. The dry canal was treated and the bulk of the herbicide residues was transported from the plots by water released through the canal in less than an hour. The herbicide was effectively reduced as the water flowed over the canal for about two miles downstream. Small detectable quantities of dichlobenil were found for a seven-day period at the downstream border of the treated area. Since these detectable residues were evident for two miles below the treated zone, the immediate reduction in herbicide content probably was due to the loss of herbicide-bearing water to the dry soil of the canal bottom. Uptake and possible metabolism by plants was not mentioned. However, Franks and Comes (1967), dealing with pond water with dense weed growth, suggested that a large fraction of the herbicide may be held by the weeds for some time. Although diquat and paraquat residues persisted for short periods in water, considerable quantities were associated with the hydrosoil which acted as an adsorbent. The lapse in time between detectability in soil and loss of level of detectability in the water was thought to be associated with absorption into weeds. Upon decay of dead weeds, the herbicides were released to the adsorbing hydrosoil or remained tightly bound to the settling organic matter from the plants.

Aquatic plants which remove a chemical by adsorption followed by possible degradation conceivably could provide a practical means of alleviating the residue problem of a particular herbicide. There are known cases of resistant terrestrial plants which are able to degrade herbicides to an inactive form. This suggests that inactivation of herbicides by resistant aquatic plants could be an important factor in the removal of the herbicide from the aquatic environment. If the level of treatment is too low, sometimes susceptible plants are inadequately controlled and may account for metabolism of the chemical. Even at sufficient dosages for control of susceptible aquatic plants, absorption would reduce the concentration of the chemical in the water; however, the herbicide may be released by decaying vegetation.

Submersed waterstargrass absorbed and translocated simazine by both leaf and root routes of entry (Funderburk and Lawrence 1963). Sutton et al. (1969) determined the removal of simazine from water by three aquatic plants (Table III). Emersed parrotfeather and elodea effectively removed simazine from the water. Using simazine-^{14}C, greater accumulations of simazine and/or products were found in the upper part of the shoots than in the roots of emersed parrotfeather plants (Table IV). The simazine-treated parrotfeather exhibited an increase in transpiration. A continuous removal of simazine from the solution combined with evidence for accumulation in the top of the shoot indicated apoplastic translocation and a possible mechanism of removal from water.

One rather obvious answer to herbicide residue problems in soils is to deliberately include crop plants that are resistant to the herbicide. This serves the twofold purpose of permitting land to stay in production while at the same time hastening the dissipation of the herbicide. Herbicides may be lost from soils by absorption into crop plants which may theoretically be used as a trap crop. For example, corn can absorb simazine and atrazine from soils and metabolize them to non-phytoxic materials (Montgomery and Freed 1961, Hamilton and Moreland 1962). Susceptible crops have a limited capacity to remove herbicide residues from soil. The role of corn, sorghum, and johnson-

Table III. *Simazine remaining in nutrient solution determined at daily intervals after treatment of various acquatic plants* (Sutton et al. 1969)

Plant	Days after treatment	Simazine in nutrient solution (p.p.m.w.) Simazine application (p.p.m.w.) [a]		
		0.12	0.50	1.00
Common duckweed	1	0.14a	0.50a	0.92a
	2	0.12a	0.41a	0.80a
	3	0.12a	0.40a	0.87a
	4	0.20a	0.50a	0.76a
Elodea	1	0.12a	0.47a	0.97a
	2	0.14a	0.39a	0.77b
	3	0.14a	0.39a	0.82b
	4	0.14a	0.47a	0.80b
Parrotfeather [b]	1	0.24a	0.50a	0.97a
	2	0.16a	0.40b	0.80b
	3	0.20a	0.34b	0.73b
	4	0.30a	0.38b	0.68b
Control	1	0.15a	0.47a	1.00a
	2	0.16a	0.49a	1.10a
	3	0.10a	0.48a	1.00a
	4	0.09a	0.42a	0.93a

[a] Values in a column within a species or the control followed by a common letter are not significantly different at the five percent level, as determined by Duncan's multiple range analysis.
[b] Emersed stage of growth.

Table IV. *Assay of emersed parrotfeather plants for radioactivity four days after root treatment with simazine-[14] C* (Sutton et al. 1969)

Plant section	Residue (c.p.m./mg. dry wt.) [a] Simazine conc. (p.p.m.w.)			
	0	0.12	0.50	1.00
Shoot				
Upper one-half	2a	37a	189a	370a
Lower one-half	1a	41a	178a	341a
Root	1a	61a	110b	188b

[a] Values in a column followed by the same letter are not significantly different as determined at the five percent level by Duncan's multiple range analysis.

grass in dissipation of atrazine from soil has been studied (Sikka and Davis 1966); all three crops were effective and about 25 percent of the initial herbicide could be accounted for by uptake into some of the plants (Table V). Other similar examples could be mentioned.

Table V. *Atrazine content in ppmw of pot soil one and 90 days after treatment* (Sikka and Davis 1966)

Treatment		Residue (p.p.m.w.) after [a,b]	
Applied (p.p.m.w.)	Crop	1 day	90 days
1	None	0.92 c	0.28 c
1	Corn	0.90 c	0.03 c
1	Johnsongrass	0.88 c	0.03 c
1	Sorghum	0.86 c	0.07 c
2	None	1.80 d	0.47 b
2	Corn	1.70 d	0.09 c
2	Johnsongrass	1.76 d	0.06 c
2	Sorghum	1.80 d	0.15 d
4	None	3.50 e	1.04 a
4	Corn	3.57 e	0.17 d
4	Johnsongrass	3.60 e	0.16 d
4	Sorghum	3.62 e	0.31 c

[a] Each value is an average of four replications.
[b] Values followed by a common letter in the same column do not differ significantly at the five percent level.

6. Modification of herbicide metabolism in soils and plants. — As discussed earlier, herbicides are subject to dissipation, in part, by chemical breakdown, microbial decomposition, and metabolic degradation within higher plants.

In almost all cases carefully studied thus far, microbial activity plays an important or even dominant role in the degradation of herbicides in soils [Kearney and Kaufman (In press)]. This principle is already used, indirectly, in removing herbicide residues from cropland. For example, tillage practices, irrigation schemes and crop rotations that provide a continuous source of organic matter in the soil are often followed which tend to optimize microbial activity and herbicide degradation. Such factors can be manipulated to some degree on a practical field scale, whereas others such as temperature cannot.

Conceivably, one might obtain the benefits of a soil-applied herbicide for a desired period then apply an inoculum of microorganisms that are specially suited to degrade that herbicide or family of compounds in question (Audus 1951 and others). For example, by continuous perfusion of aerated solutions of 2,4-D, MCPA, and 2,4,5-T through garden soil, AUDUS (1951) found indications that the kinetics of breakdown of these herbicides were essentially similar. Evidence for the development of a bacterial (or mixed microbiological) flora adapted to herbicide breakdown was thus strongly suggested. Although this approach has been considered for many years by a number of investigators, it has had no practical acceptance yet (to our knowledge), because of numerous uncontrolled ecological and other factors which themselves require further study.

A few deliberate attempts at direct chemical inactivation of herbicides in the soil have been made with some promise. Corn is known to possess a natural sweet substance (the 2-glucoside of 2,4-dihydroxy-7-methoxy-1, 4-benzoxazin-3-one) that is capable of detoxifying several 2-chloro-substituted s-triazines by converting them to their 2-hydroxy analogs (CASTELFRANCO et al. 1961, MONTGOMERY and FREED 1961, HAMILTON and MORELAND 1962, PALMER and GROGAN 1965). CASTELFRANCO and BROWN (1962) also found that pyridine and hydroxylamine brought about the conversion of simazine-C^{14} to the 2-hydroxy analog. Based on these substances as models, a mechanism for the dechlorination of simazine in *Zea* was proposed. Since the detoxification was non-enzymatic, CASTELFRANCO and DEUTSCH (1962) reasoned that similar reactions might be used to detoxify simazine after it has been applied to soil. They found that aqueous solutions of sodium and calcium polysulfides converted C^{14}-labeled simazine, probably to hydroxy-simazine, *in vitro*. In soil, calcium polysulfide accelerated the breakdown of simazine as determined by oat bioassay. These results suggest the possibility of using polysulfides to remove unwanted residues of chlorotriazine herbicides from croplands.

Potential herbicidal antidotes have also been investigated as chemical seed treatments (HOFFMAN 1962). Approximately 75 percent of the inhibitory effect of a foliar application of barban on wheat was eliminated by dusting the seed before planting with 4'-chloro-2-hydroxyiminoacetanilide at a rate of one oz./bushel. Other highly effective antidotes were 2,4-dichloro-9-xanthenone and N-methyl-3,4-dichlorobenzene sulfonamide. Among several compounds studied, barban antidotes were found in those chemicals that produced formative effects on plants. Although the mechanism of action whereby chemical antidotes are able to mitigate herbicidal injury on crop plants is unknown, this concept introduces some exciting new prospects in plant growth regulation as well as offering possible promise for minimizing herbicide residues.

Conceivably, by including an appropriate chemical additive (either adjuvant or antidote) along with the herbicide, or else introducing such additives separately before or after herbicide application, one might actually be able to control the metabolic degradation of the herbicide within plant tissue. A promising recent example is cited in support of this contention [SMITH et al. (1968)]. Several factors were studied which influenced amitrole metabolism in excised leaves of bean (*Phaseolus vulgaris* L. var. Red Kidney) and Canada thistle (*Circium arvense* (L.) Scop.) ecotypes. Relevant to this discussion were the effects of the chemical additives ammonium thiocyanate and benzyladenine. Dipping the leaves in $10^{-2}M$ ammonium thiocyanate reduced the metabolism of amitrole, whereas similar pre-treatment with $3.5 \times 10^{-3}M$ benzyladenine enhanced the metabolic degradation of amitrole (Table VI). The results are consistent with field experience which often shows amitrole plus ammonium thiocyanate to be herbicidally more effective than amitrole alone.

CARTER (1965) also observed that ammonium thiocyanate at $10^{-2}M$ decreased the metabolism of amitrole in bean leaves and HERRETT and LINCK (1961) showed that sodium fluoride at $10^{-1}M$ inhibited amitrole metabolism in Canada thistle and field bindweed (*Convolvulus arvensis* L.).

Controlled metabolism of herbicides within plant tissue could have far-reaching implications for residue accumulation or removal and for the field of plant growth regulation generally.

Summary

Weed science is a rapidly expanding new discipline. About 120 herbicides now account for approximately 55 percent of the total dollar value of pesticides in the United States; the rate of increase of herbicide use in recent years exceeds that of all other groups of pesticides. The impact of weeds and the increasing use of chemical herbicides is felt by virtually every segment of the population, and the possible occurrence of herbicide residues in the environment is naturally of public concern. The importance of knowing and being able to manipulate the distributions and fates of these compounds in the biosphere is apparent.

Three principal approaches to minimize residue accumulation of pesticide chemicals, including herbicides are: (a) use an alternative (non-chemical) method; (b) enhance the toxicity, selectivity, or effectiveness of chemical methods to require less pesticide and (c) effect dissipation of the residue once contamination has occurred. The first two alternatives are preventive, the last corrective. Several research approaches promise to alleviate the herbicide residue situation: (1) use of non-chemical means of vegetation control; (2) reduction of dosages by increasing effectiveness through (a) optimum placement for foliar and/or root uptake, (b) reduced volatilization, (c) enhanced penetration and translocation, (d) modification of herbicide metabolism in plants and soils, (e) alteration of mobility, availability, and residual life in soils; (3) use of specific trap plants to remove them from water and soils; and (4) use of adsorbents to remove or inactivate residues in water, air, and soils.

Table VI. *The relative amounts of amitrole-*[14]* C, unknown I, and unknown II in excised bean and Canada thistle leaves as influenced by certain chemical treatments* [SMITH *et al.* (1968)]

Treatment	Total ^{14}C recovered on chromatograms (%)			
	Amitrole-[14]C	Unknown II	Unknown I	Total
Untreated				
Bean	24.3	51.3	11.9	87.5
Canada thistle				
ecotype				
YM	53.5	20.9	14.8	89.2
G1	45.1	23.5	19.9	88.5
FI	36.9	19.9	29.8	86.6
Tween 20				
(1 g./l.)				
Bean	18.6	56.7	12.8	88.1
Canada thistle				
ecotype				
YM	42.3	24.9	24.3	91.5
G1	42.8	20.7	24.7	88.2
FI	38.5	19.6	28.1	86.2
N^6-benzyladenine				
$(3.5 \times 10^{-2}M)$				
Bean	14.6	60.0	10.9	85.5
Canada thistle				
ecotype				
YM	51.0	31.8	7.3	90.1
G1	44.8	22.5	23.5	90.8
FI	31.0	21.3	36.5	89.2
N^6-benzyladenine				
$(3.5 \times 10^{-3}M)$				
Bean	14.6	50.5	21.6	86.7
Canada thistle				
ecotype				
YM	41.3	25.6	19.7	86.8
G1	29.3	20.9	37.2	87.9
FI	12.4	23.7	48.0	84.1
Ammonium thiocyanate				
$(10^{-2}M)$				
Bean	74.8	12.0	4.2	91.0
Canada thistle				
ecotype				
YM	75.5	9.3	6.0	90.8
G1	72.1	11.8	7.8	91.7
FI	65.2	9.4	13.1	87.7
Ammonium thiocyanate				
$(10^{-3}M)$				
Bean	46.4	27.8	6.4	90.6
Canada thistle				
ecotype				
YM	47.6	19.5	21.8	88.9
G1	44.0	19.0	23.7	86.7
FI	30.5	13.2	44.7	88.4

Table VII. *Chemical names of herbicides and other compounds mentioned by common name in the text*

Common name	Chemical name
acrolein	Acrylaldehyde
Aliquat 204	Dilauryl dimethyl ammonium chloride
Aliquat 221	Dicoco dimethyl ammonium chloride
ametryne	2-(Ethylamino)-4-(isopropylamino)-6-(methylthio)-s-triazine
amiben	3-Amino-2,5-dichlorobenzoic acid
ammonium thiocyanate	NH$_4$CN
atrazine	2-Chloro-4-(ethylamino)-6-(isopropylamino)-s-triazine
barban	4-Chloro-2-butynyl m-chlorocarbanilate
bensulide	O,O-Diisopropyl phosphorodithioate S-ester with N-(2-mercaptoethyl)benzenesulfonamide
benzyladenine	N^6-Benzyladenine
bromacil	5-Bromo-3-sec-butyl-6-methyluracil
CDEC	2-Chloroallyl diethyldithiocarbamate
chloropicrin	Trichloronitromethane
chloroxuron	3-[p-(p-chlorophenoxy)phenyl]-1,1-dimethyl-urea
CIPC (Chlorpropham)	Isopropyl m-chlorocarbanilate
dalapon	2,2-Dichloropropionic acid
2,4-D	(2,4-Dichlorophenoxy(acetic acid
2,4-DB	4-(2,4-Dichlorophenoxy)butyric acid
DDT	1,1,1-Trichloro-2,2-bis(p-chlorophenyl)ethane
2,4-DEB	2-(2,4-Dichlorophenoxy)ethyl benzoate
2,4-DEP	Tris[2-(2,4-dichlorophenoxy)ethyl] phosphite
dicamba	3,6-Dichloro-o-anisic acid
dichlobenil	2,6-Dichlorobenzonitrile
dipropalin	N,N-Dipropyl-2,6-dinitro-p-toluidine
diquat	6,7-Dihydrodipyrido[1,2-a:2′,1′-c]pyrazidiinium salts
diuron	3-(3,4-Dichlorophenyl)-1,1-dimethylurea
endothall	7-Oxabicyclo[2.2.1]heptane-2,3-dicarboxylic acid
EPTC	s-Ethyl dipropylthiocarbamate
fenac	(2,3,6-Trichlorophenyl) acetic acid
fenuron	1,1-Dimethyl-3-phenylurea
IPC (propham)	Isopropyl carbanilate
isocil	5-Bromo-3-isopropyl-6-methyluracil
linuron	3-(3,4-Dichlorophenyl)-1-methoxy-1-methylurea
MCPA	[(4-Chloro-o-tolyl)oxy] acetic acid
MCPB	4-[(4-Chloro-o-tolyl)oxy] butyric acid
MCPES	2-[(4-Chloro-o-tolyl)oxy] ethyl sodium sulfate
methyl bromide	CH$_3$Br
monuron	3-(p-Chlorophenyl)-1,1-dimethylurea
MOPA	alpha-Methoxyphenylacetic acid
neburon	1-Butyl-3-(3,4-dichlorophenyl)-1-methylurea
paraquat	1,1′-Dimethyl-4,4′-bipyridinium salts
PBA	Mixed isomers containing 2,3,6-TBA and 2,3,5,6-tetrachlorobenzoic acid
pebulate	S-Propyl butylethylthiocarbamate
picloram	4-Amino-3,5,6-trichloropicolinic acid
simazine	2-Chloro-4,6-bis(ethylamino)-s-triazine
sodium fluoride	NaF

Table VII. (Continued)

Common name	Chemical name
trifluralin	a,a,a-Trifluoro-2,6-dinitro-N,N-dipropyl-p-toluidine
2,4,5-T	(2,4,5-Trichlorophenoxy) acetic acid
2,3,5,6-TBA	2,3,5,6-Tetrachlorobenzoic acid
2,3,6-TBA	2,3,6-Trichlorobenzoic acid
2,4,5-TES	Sodium 2-(2,4,5-trichlorophenoxy) ethyl sulfate

Organic pesticides in air, water, plants, and soil are subject to degradation, alteration of structure, or conjugation with natural constituents. Virtually all organic herbicides known thus far are eventually dissipated in nature. Whereas many of the herbicides now in common use are relatively short lived and innocuous, the distribution and fate of others are less understood. The more persistent ones deserve further attention because of possible cycling in the biosphere and accumulation in air, water, feed products, and man's food. Publications dealing with purposeful attempts at herbicide decontamination in the biosphere are limited. Most emphasis thus far has been placed on avoiding or minimizing residues.

Résumé *

Quelques thèmes de recherches pour la réduction des résidus d'herbicides dans le milieu environnant

La science des plantes vivaces est une discipline nouvelle en rapide expansion. L'exploitation de près de 120 herbicides représente actuellement environ 40 pour cent du chiffre d'affaires des pesticides aux Etats-Unis ; le taux de croissance de l'emploi des herbicides, ces dernières années, dépasse celui de tous les autres groupes de pesticides. L'incidence des mauvaises herbes et de l'emploi en augmentation des herbicides chimiques est ressentie pratiquement par tout élément de la population et la présence possible de résidus d'herbicide dans le milieu environnant est naturellement d'intérêt public. Connaître la distribution et le sort de ces composés dans la biosphère et avoir la possibilité d'intervenir, apparaît très important. Il est trois principaux moyens pour réduire l'accumulation des résidus de pesticides chimiques, y compris les herbicides ; a) l'emploi d'une autre méthode sans produits chimiques, b) exalter la toxicité, la sélectivité ou l'efficacité des méthodes chimiques de manière à utiliser moins de pesticides et, c) produire la dispersion du résidu lorsque la pollution est survenue. Les deux premières méthodes sont préventives, la dernière corrective. Plusieurs idées de recherche promettent de soulager la situation des résidus d'herbicides : (1) emploi des moyens non chimiques de con-

---

* Traduit par R. MESTRES.

trôle de la végétation, (2) réduction des doses par augmentation de l'efficacité grâce à : (a) une application optimale pour l'absorption par les feuilles ou les racines, (b) une volatilisation réduite, (c) une pénétration et un transfert amélioré, (d) une modification du métabolisme des herbicides dans les plantes et les sols, (e) altération de la mobilité et de la remanence dans les sols — (3) l'emploi d'outils spécifiques pour retirer les plantes de l'eau et des sols ; et (4) emploi d'adsorbants pour enlever ou inactiver les résidus dans l'eau, l'air et les sols. Les pesticides organiques dans l'air, l'eau, les plantes et les sols sont sujets à la dégradation, l'altération de structure ou la conjugaison avec des composés naturels. Tous les herbicides organiques connus jusqu'ici sont finalement dissipés dans la nature. Alors que beaucoup d'herbicides actuellement d'usage courant ont une rémanence faible et sont inoffensifs, la distribution et le sort des autres sont moins compris. Les plus persistants méritent plus d'attention à cause d'un cycle possible dans la biosphère et une accumulation dans l'air, l'eau, les aliments pour le bétail et la nourriture de l'homme. Les publications ayant trait à des tentatives de décontamination des herbicides de la biosphère sont en nombre limité. Il a été surtout question jusqu'ici d'éviter ou de réduire les résidus.

Zusammenfassung[*]

Einige Forschungsversuche zur Verringerung von Herbizidrückständen in der Umgebung

Unkrautkenntnis ist eine schnell wachsende neue Wissenschaft. Etwa 120 Herbizide representieren heute ungefähr 40 prozent des Gesamtdollarwertes der Pestizide in den Vereinigten Staaten. Die Zuwachsrate an Herbizidverbrauch in den letztten Jahren übertrifft die aller andern Pestizidgruppen. Der Einfluss des Unkrauts und der steigende Gebrauch von chemischen Herbiziden wird von eigentlich jedem Teil der Bevölkerung gespürt, und das mögliche Auftreten von Herbizidrückständen in der Umgebung ist natürlicherweise von allgemeinem Belang. Die Wichtigkeit, die Verteilungen und Schicksale dieser Verbindungen in der Biosphäre zu kennen und fähig zu sein, sie zu handhaben, ist einleuchtend.

Drei grundsätzliche Methoden, um Rückstandsakkumulation von Pestizidchemikalien, einschliesslich Herbiziden, zu verringern sind (a) der Gebrauch einer anderen (nicht chemischen) Methode, (b) die Toxizität, Selektivität oder Wirksamkeit von chemischen Methoden so zu erweitern, dass weniger Pestizid erforderlich ist und (c) die Vertreibung der Rückstände zu bewirken, wenn einmal Kontamination eingetreten ist. Die ersten zwei Möglichkeiten verhindern, die letzte

---

[*] Übersetzt von A. Schumann.

korrigiert. Mehrere Forschungsversuche haben Aussicht, die Herbizid-rückstandssituation zu erleichtern: (1) der Gebrauch von nichtche-mischen Mitteln zur Unkrautkontrollierung, (2) Reduktion der An-wendungsmengen durch steigende Wirksamkeit durch (a) optimale Plazierung für Blatt- und/oder Wurzelaufnahme, (b) reduzierte Flüchtigkeit, (c) gesteigerte Durchdringung und Verlagerung, (d) Modifizierung des Herbizidmetabolismus in Pflanzen und im Boden, (e) Aenderung der Beweglichkeit, Verfügbarkeit und Rückstands-lebensdauer im Boden; (3) der Gebrauch von spezifischen Fangpflan-zen, um die Herbizide aus Wasser und boden zu entfernen und (4) der Gebrauch von Absorptionsmitteln, um Rückstände in Wasser, Luft und Boden zu entfernen oder zu inaktivieren.

Organische Pestizide in Luft, Wasser, Pflanzen und Boden sind dem Abbau, der Strukturveränderung oder Verbindung mit natürlichen Bestandteilen ausgesetzt. Alle bisher bekannten organischen Pestizide werden schliesslich in der Natur vernichtet. Während viele der zurzeit gebräuchlichen Herbizide relativ kurzlebig und harmlos sind, sind die Verteilungen und Schicksale von anderen weniger verstanden. Die beständigeren Chemikalien verdienen weitere Aufmerksamkeit wegen möglichen Kreislaufs in der Biosphäre und Anhäufung in Luft, Wasser, Futtermitteln und menschlichen Nahrungsmitteln. Veröffentlichungen, welche sich mit zweckmässigen Versuchen zur Herbiziddekontamina-tion in der Biosphäre befassen, sind begrenzt. Die Hauptbetonung hat bisher darauf gelegen, Rückstände zu vermeiden oder zu verringern.

References

AHRENS, J. F.: Detoxification of simazine and atrazine treated soil with activated carbon. Proc. N.E. Weed Control Conf. 19, 364 (1965 a).
—— Improving herbicide selectivity in horticultural crops with activated carbon. Proc. N.E. Weed Control Conf. 19, 366 (1965 b).
ALEXANDER, M., and M. I. H. ALEIM: Effect of chemical structure on microbial decomposition of aromatic herbicides. J. Agr. Food Chem. 9, 44 (1961).
ANDERSON, A. H.: The inactivation of simazine and linuron in soil by charcoal. Weed Research 8, 58 (1968).
ARLE, H. F., O. A. LEONARD, and V. C. HARRIS: Inactivation of 2,4-D on sweet potato slips with activated charcoal. Science 107, 247 (1948).
ASHTON, F. M., D. A. HARVEY, and C. L. FOY: Principles of selective weed control Calif. Agr. Expt. Sta. Extension Serv. Circ. 505, (1961).
AUDUS, L. J.: The biological detoxication of hormone herbicides. Plant and Soil 3, 170 (1951).
—— (ed.): The physiology and biochemistry of herbicides. London-New York: Academic Press (1964).
AYA, F. O., and S. K. RIES: Influence of oils on the toxicity of amitrole to quack-grass. Weed Sci. 16, 288 (1968).
BANDEEN, J. D., and W. C. VERSTRAETE: Effects of several oils and surfactants on the enhancement of atrazine for weed control in corn. Weed Soc. Amer. (Abstr.), p. 9 (1967).
BAYER, D. E.: Effects of surfactants on leaching of substituted urea herbicides in soil. Weeds 15, 249 (1967).

Bingham, S. W.: Residue of bensulide in turfgrass soil following annual treatments for crabgrass control. Agron. J. 59, 327 (1967).

——, and R. E. Schmidt: Crabgrass control in turf. Proc. S. Weed Conf. 17, 113 (1964).

Burnside, O. C., G. A. Wicks, and C. R. Fenster: The effect of rainfall and soil type on the disappearance of 2,3,6-TBA. Weeds 11, 45 (1963).

Carter, M.: Studies on the metabolic activity of 3-amino-1,2,4-triazole. Physiol. Plant. 18, 1054 (1965).

Castelfranco, P., and M. S. Brown: Purification and properties of the simazine-resistance factor of Zea mays. Weeds 10, 131 (1962).

——, and D. Deutsch: Action of polysulfide ion on simazine in soil. Weeds 10, 244 (1962).

——, C. L. Foy, and D. Deutsch: Non-enzymatic detoxification of simazine by extracts of Zea mays. Weeds 9, 580 (1961).

Coats, G. E., and C. L. Foy: Effect of Tween 80 and DMSO on the absorption and translocation of three phloem-mobile herbicides in Verbascum thapsus L. Assoc. S.E. Biol. Bull. 15, (2), 34 (1968).

Comes, R. D., and F. L. Timmons: Effect of sunlight on the phytotoxicity of some phenylurea and triazine herbicides on a soil surface. Weeds 13, 81 (1965).

Corbin, F. T., and R. P. Upchurch: Influence of pH on detoxication of herbicides in soil. Weed Sci. 15, 370 (1967).

Crafts, A. S., and C. L. Foy: The physical and chemical nature of plant surfaces in relation to the use of plant surfaces and their residues. Residue Reviews, 1, 112 (1962).

——, and W. W. Robbins: Weed control, 3rd ed. New York: McGraw-Hill (1962).

Currrier, H. B., and C. D. Dybing: Foliar penetration of herbicides—Review and present status. Weeds 7, 195 (1959).

Darding, R. L., and J. F. Freeman: Residual toxicity of fluometuron in soils. Weed Sci. 16, 266 (1968).

Day, B. E.: Formulation of dalapon to reduce soil contamination. Weed Research 1, 177 (1961).

——, L. S. Jordan, and V. A. Jolliffe: The influence of soil characteristics on the adsorption and phytotoxicity of simazine. Weed Sci. 16, 209 (1968).

—— ——, and R. C. Russell: Persistence of dalapon residues in California soils. Soil Sci. 95, 326 (1963).

Donaldson, T. W., and C. L. Foy: The phytotoxicity and persistence in soils of benzoic acid herbicides. Weeds 13, 195 (1965).

Ebeling, W.: Analysis of the basic processes involved in the deposition, degradation, persistence, and effectiveness of pesticides. Residue Reviews 3, 35 (1963).

Foy, C. L.: Foliar penetration—Review of herbicide penetration through plant surfaces. J. Agr. Food Chem. 12, 473 (1964).

——, and G. E. Coats: Results 1968 field research program. Plant Pathol. Physiol. Information Note 108, (1968).

——, and L. W. Smith: Surface tension lowering, wettability of paraffin and corn leaf surfaces, and herbicidal enhancement of dalapon by seven surfactants. Weeds 13, 15 (1965).

—— —— The role of surfactants in modifying the activity of herbicidal sprays. Adv. Chem. Series, In press (1968).

——, J. W. Whitworth, T. J. Muzik, and H. B. Currier: The penetration, absorption, and translocation of herbicides. In: Environmental and other factors in the response of plants to herbicides. Ore. State Univ. Agr. Expt. Sta. Tech. Bull. No. 100 (1967).

FRANKS, P. A., and R. D. COMES: Herbicide residues in pond water and hydrosoil. Weeds **15**, 210 (1967).

——, R. H. HODGSON, and R. D. COMES: Residues of two herbicides in water in irrigation canals. Weeds **15**, 353 (1967).

FUNDERBURK, H. H., JR., and J. M. LAWRENCE: Preliminary studies on the absorption of C^{14}-labeled herbicides in fish and aquatic plants. Weeds **11**, 217 (1963).

——, N. S. NEGI, and J. M. LAWRENCE: Photochemical decomposition of diquat and paraquat. Weeds **14**, 240 (1966).

GRAY, R. A.: A vapor trapping apparatus for determining the loss of EPTC and other herbicides from soils. Weeds **13**, 138 (1965).

——, and A. J. WEIERICH: Factors affecting the vapor loss of EPTC from soils. Weeds **13**, 141 (1965).

GRZENDA, A. R., H. P. NICHOLSON, and W. S. COX: Persistence of four herbicides in pond water. J. Amer. Water Works Assoc. **58**, 326 (1966).

HAMILTON, R. H., and D. E. MORELAND: Simazine degradation by corn seedlings. Science **135**, 373 (1962).

HANCE, R. J.: Observations on the relationship between the adsorption of diuron and the nature of the adsorbent. Weed Research **5**, 108 (1965).

HARRIS, C. I.: Movement of herbicides in soil. Weeds **15**, 214 (1966).

——, and T. J. SHEETS: Influence of soil properties on adsorption and phytotoxicity of CIPC, diuron, and simazine. Weeds **13**, 215 (1965).

HERRETT, R. A., and A. J. LINCK: The metabolism of 3-amino-1,2,4-triazole by Canada thistle and field bindweed and the possible relation to its herbicidal action. Physiol. Plant. **14**, 767 (1961).

HILL, G. D., JR., I. J. BELASCO, and H. L. PHLOEG: Influence of surfactants on the activity of diuron, linuron, and bromacil foliar sprays on weeds. Weeds **13**, 103 (1965).

HILTON, J. L., L. L. JANSEN, and H. HULL: Mechanisms of herbicide action. Ann. Review Plant Physiol. **14**, 353 (1963).

HOFFMAN, O. L.: Chemical seed treatments as herbicidal antidotes. Weeds **10**, 322 (1962).

HOWARTH, R.: Grower finds advantages in wilting agent. S. Florist and Nurseryman **74** (47), 12 (1962).

HURTT, W., and C. L. FOY: Excretion of foliarly-applied dicamba and picloram from roots of Black Valentine beans grown in soil, sand and culture solution. Proc. N.E. Weed Control Conf. **19**, 602 (1965 a).

—— —— Some factors influencing the excretion of foliarly-applied dicamba and picloram from roots of Black Valentine beans. Plant Physiol. (Suppl.) **40**, xlviii (1965 b).

HYNDSHAW, A. Y.: Treatment application points for activated carbon. Taste Odor Control J. **28** (3), 1 (1962).

JAZSCHITZ, J. A.: Use of charcoal to deactivate herbicide residues in turfgrass seedbeds. Proc. N.E. Weed Control Conf. **22**, 401 (1968).

——, and C. R. SKOGLEY: Turfgrass response to seedbed and seedling applications of peremergence and broadleaf herbicides. Proc. N.E. Weed Control Conf. **20**, 554 (1966).

JANSEN, L. L.: Surfactant enhancement of herbicide entry. Weeds **12**, 251 (1964 a).

—— Relation of structure of ethylene oxide ether type nonionic surfactants to herbicidal activity of water soluble herbicides. J. Agr. Food Chem. **12**, 223 (1964 b).

—— Herbicidal and surfactant properties of long-chain alkylamine salts of 2,4-D in water and oil sprays. Weeds **13**, 123 (1965 a).

—— Effects of structural variations in nonionic surfactants of phytotoxicity and physical-chemical properties of aqueous sprays of several herbicides. Weeds 13, 117 (1965 b).

——, W. A. Gentner, and W. C. Shaw: Effects of surfactants on the herbicidal activity of several herbicides in aqueous spray systems. Weeds 9, 381 (1961).

Jones, D., and C. L. Foy: Tracer studies with three ^{14}C-labeled herbicides, DMSO and Tween 80 in Black Valentine bean. Assoc. S.E. Biol. Bull. 15 (2), 42 (1968).

Jordan, L. S., B. E. Day, and W. A. Clerx: Photodecomposition of triazines. Weeds 12, 5 (1964).

——, J. D. Mann, and B. E. Day: Effects of ultraviolet light on herbicides. Weeds 13, 43 (1965).

Kearney, P. C.: Metabolism of herbicides in soils. In: Organic pesticides in the environment. Adv. Chem. Series 60, 250 (1966).

——, and D. D. Kaufman ed.): Degradation of herbicides. New York: Marcel Dekker (1969).

Klingman, G. C.: Weed control: As a science. New York: Wiley (1961).

Koren, E., C. L. Foy, and F. M. Ashton: Relative adsorption and migration of four thiolcarbamates in five soil types. Weed Sci., In press (1968).

Lange, A. H.: Problems of soil contamination, accumulation of residues and the effect on subsequent croppings, Proc. W. Weed Control Conf., p. 30 (1967).

Linder, P. W., J. C. Craig, Jr., F. E. Cooper, and J. W. Mitchell: Movement of 2,3,6-trichlorobenzoic acid from one plant to another through their root systems. J. Agr. Food Chem. 6, 356 (1958).

——, J. W. Mitchell, and G. D. Freeman: Persistence and translocation of exogenous regulating compounds that exude from roots. J. Agr. Food Chem. 12, 437 (1964).

Linscott, D. L., and R. D. Hagin: Protecting alfalfa seedlings from a triazine with activated charcoal. Weed Sci. 15, 304 (1967).

Lucas, E. H., and C. L. Hammer: Inactivation of 2,4-D by adsorption on charcoal. Science 105, 340 (1967).

Mahan, J. N., D. L. Fowler, and H. H. Shepard: The pesticide review—1968. The Pesticide Review, Dec. (1968).

McCutcheon, J. W.: Detergents and emulsifiers (Annual). Morristown, N.J.: McCutcheon (1964).

McRae, I. C., and M. Alexander: Microbial degradation of selected herbicides in soil. J. Agr. Food Chem. 13, 72 (1965).

McWhorter, C. G.: Effects of surfactant concentration on Johnsongrass control with dalapon. Weeds 11, 83 (1963).

Mitchell, J. W., B. C. Smale, and W. H. Preston, Jr.: New plant regulators that exude from roots. J. Agr. Food Chem. 7, 841 (1959).

Montgomery, M., and V. H. Freed: The uptake, translocation, and metabolism of simazine and atrazine by corn plants. Weeds 9, 231 (1961).

Moreland, D. E.: Mechanisms of action of herbicides. Ann. Review Plant Physiol. 18, 365 (1967).

Palmer, R. P., and C. O. Grogan: Tolerance of corn lines to atrazine in relation to content of benzoxazinone derivative, 2-glucoside. Weeds 13, 219 (1965).

Parochetti, J. V., and G. F. Warren: Vapor losses of IPC and CIPC. Weeds 14, 281 (1966).

Pelishek, R. E., J. Osborn, and J. Letey: The effect of wetting agents on infiltration. Proc. Soil Sci. Soc. Amer. 26, 595 (1962).

Pesticide Regulation Division, U.S. Department of Agriculture (ARS): Summary of registered agricultural pesticide chemical uses—With supplements (1968).

Preston, W. H., J. W. Mitchell, and W. Reeve: Movement of alpha-methoxyphenylacetic acid from one plant to another through their root systems. Science **119**, 437 (1954).

RODGERS, E. G.: Leaching of seven *s*-triazines. Weed Sci. **16**, 117 (1968).

SCHECHTER, M. S.: Editorial—The need for confirmation. Pesticides Monitoring J. **2** (1), 1 (1968).

SCHUBERT, O. E.: Can activated charcoal protect crops from herbicide injury? Crops and Soils **19** (9), 10 (1967).

SHAW, W. C., J. L. HILTON, D. E. MORELAND, and L. L. JANSEN: Herbicides in plants. *In:* The nature and fate of chemicals applied to soils, plants, and animals. *U.S. Department of Agriculture,* ARS 20-9, p. 119 (1960).

SHEETS, T. J.: Photochemical alteration and inactivation of amiben. Weeds **11**, 186 (1963).

——, A. S. CRAFTS, and H. R. DREVER: Influence of soil properties on the phytotoxicities of the *s*-triazine herbicides. J. Agr. Food Chem. **10**, 458 (1962).

SIKKA, H. C., and D. E. DAVIS: Dissipation of atrazine from soil by corn, sorghum and johnsongrass. Weeds **14**, 289 (1966).

SMITH, L. W., and D. E. BAYER: Soil adsorption of diuron as influenced by surfactants. Soil Sci. **103**, 328 (1967).

——, D. E. BAYER, and C. L. FOY: Influence of environmental and chemical factors on amitrole metabolism in excised leaves. Weed Sci., **16**, 527 (1968).

——, and C. L. FOY: The possible mode of action of nonionic surfactants in herbicide solutions. Proc. W. Weed Control Conf., p. 139 (1966 a).

—— —— Interactions of several paraquat-surfactant mixtures. Weeds **15**, 67 (1966 b).

—— —— Penetration and distribution studies in bean, cotton, and barley from foliar and root applications of Tween 20-C^{14}, fatty acid and oxyethylene labeled. J. Agr. Food Chem. **14**, 117 (1966 c).

SUTTON, D. L., D. A. DURHAM, S. W. BINGHAM, and C. L. FOY: Influence of simazine on apparent photosynthesis of aquatic plants and herbicide residue removal from water. Weed Sci. **17**, 56 (1969).

TAYLORSON, R. B.: Decomposition of CDEC by far ultraviolet radiation. Weed Sci. **14**, 155 (1966).

UPCHURCH, R. P.: The influence of soil factors on the phytotoxicity and plant selectivity of diuron. Weeds **6**, 161 (1958).

——, J. A. KEATON, and F. L. SELMAN: Soil sterilization properties of monuron, diuron, simazine and isocil. Weed Sci. **16**, 358 (1968).

——, and D. D. MASON: The influence of soil organic matter on the phytotoxicity of herbicides. Weeds **10**, 9 (1962).

——, F. L. SELMAN, D. D. MASON, and E. J. KAMPRATH: The correlation of herbicidal activity with soil and climatic factors. Weed Sci. **14**, 42 (1966).

WOODFORD, E. K., K. HOLLY, and C. C. McCREADY: Herbicides. Ann. Review Plant Physiol. **9**, 331 (1958).

——, and G. R. SAGAR (ed.): Herbicides and the soil. Oxford: Blackwell (1960).

Decontamination of pesticides in soils

By

P. C. KEARNEY,* E. A. WOOLSON,* J. R. PLIMMER,* AND A. R. ISENSEE *

Contents

I. Introduction

The soil, as a medium for decontamination, offers a large number of processes by which organic substances can be destroyed. As such, progressive accumulation of organic pesticides would appear to be unlikely. Unfortunately, the chemical and physical properties of certain insecticides and herbicides afford them a degree of stability against the natural destructive processes in soils. The stability of these compounds is best illustrated in a recent summary of persistence data on 12 major classes of pesticides in a number of soil types (Fig. 1) (KEARNEY et al. 1969). Persistence values are expressed in months and each bar represents one or more classes of herbicide or insecticide. Each open space in the bar represents an individual pesticide falling within the larger chemical class of compounds. The length of each bar depicts the time for each class of pesticide to decrease 75 to 100 percent of the amount applied. These values are based on normal rates of application. As anticipated, the organochlorine insecticides are the most persistent pesticides. The organic herbicides persist for a few days or for more than 12 months depending on their respective properties. Only the

---

* U.S. Department of Agriculture, ARS, CR, Beltsville, Md. Specific mention of trademark instruments does not constitute an endorsement by the U.S. Department of Agriculture over others designed to give similar performances.

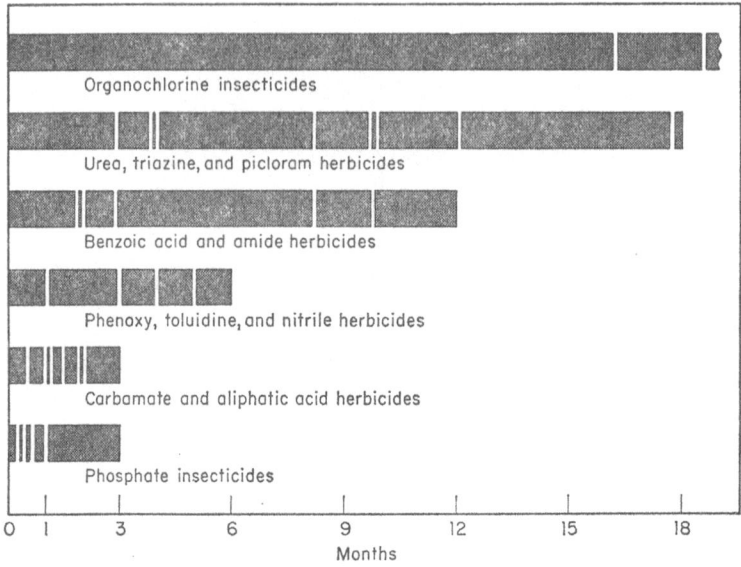

Fig. 1. Persistence of pesticides in soils

major herbicides that persist for a month or longer are shown in Figure 1. The phosphate insecticides do not persist for long periods in most soils. A more detailed picture of organochlorine pesticide persistence is shown in Figure 2. Chlordane and DDT usually persist for several years while heptachlor and aldrin extend their activity through the formation of their respective metabolites, *i.e.*, heptachlor epoxide and dieldrin.

Why are we concerned about pesticide residues in soils? Their effects would appear to be very subtle and not directly related to man or his environment. The need for pure water is obvious, for man directly consumes processed water. Not so with soil, and therefore, why is there concern over soil contamination?

There now exists unequivocal evidence that most plants can absorb and translocate residual pesticides from contaminated soils (NASH 1968). Uptake and translocation have been demonstrated with radio-labeled pesticides incorporated directly into the soil and then seeded with several agronomic crops. Many of the soil variables that influence this uptake process have been studied. For example, increasing the concentration of dieldrin and DDT in soil causes a corresponding increase in the amount of insecticide recovered by the wheat plant. The plant is apparently indiscriminate in its ability to absorb most substances from soils. Therefore a link exists between residual pesticides in soils and man's food chain. In addition, residual pesticides are potential pollutants of water and air.

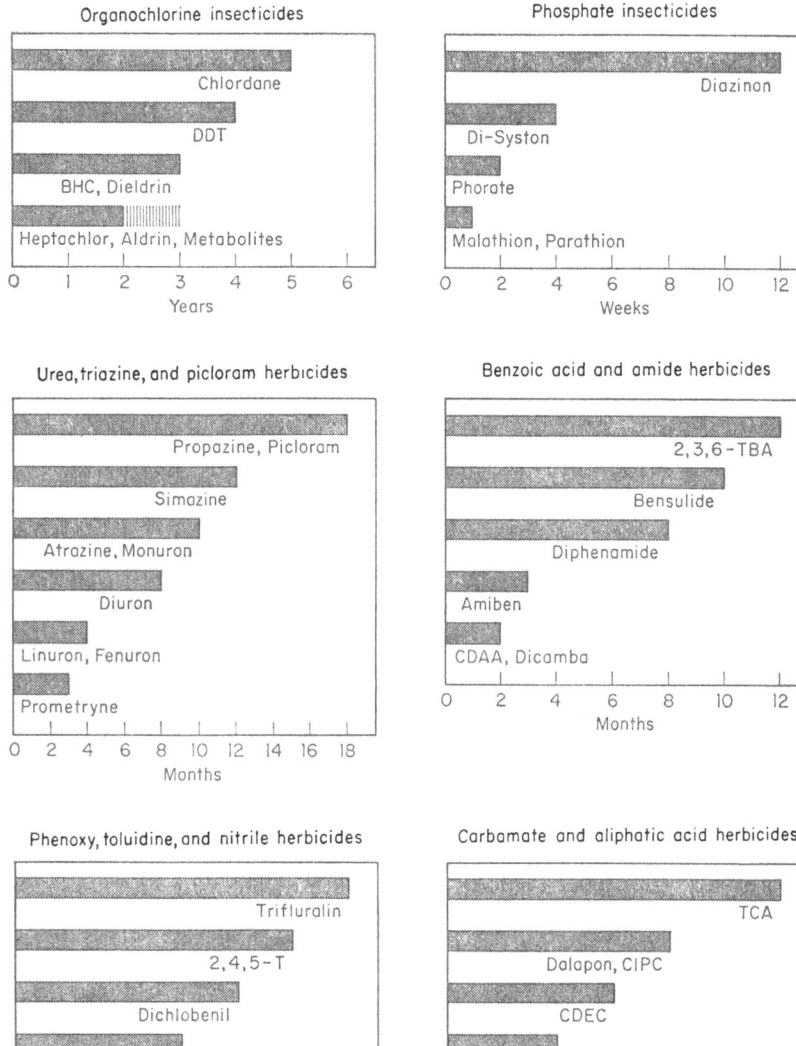

Fig. 2. Persistence of individual pesticides in soils

Now that we have defined the problem and the need for decontamination in soils, what methods are available for removing persistent pesticides? The fate of a pesticide in soils is determined by a number of processes which come under the general heading of physical, biological, and chemical (Fig. 3). Under physical, they include photo-

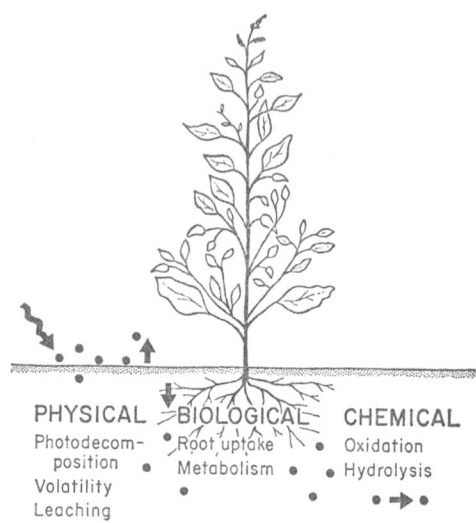

PHYSICAL ⋍BIOLOGICAL⋍ CHEMICAL
Photodecom-   • Root uptake   • Oxidation
    position •   Metabolism •   • Hydrolysis
Volatility   •   •
Leaching   • ➡ •

Fig. 3. Processes that determine fate of a pesticide in soil

decomposition, volatility, leaching, and adsorption. Under biological, they include root uptake and microbial metabolism and under chemical, they include oxidation, reduction, and hydrolysis. Several of these are responsible for decomposing pesticides. For example, soil enrichment techniques for the proliferation of specific microorganisms effective in metabolizing foreign substances have been a favorite method for microbiologists. It is conceivable that "catch plants" or plants with a high affinity for certain pesticides could be grown on contaminated soils and then removed after taking up some part of the residual pesticides. It is possible that a combination or several of these methods could be employed to reduce pesticide concentrations in soils.

If each of these processes were active on a pesticide, then soil residues would not exist. If soil microorganisms could be induced to rapidly metabolize DDT to the level of carbon dioxide, then soils would be an ideal medium for decontamination. They don't, and therein lies the problem of reducing soil residues. The work to be reported today deals with two approaches to reducing pesticide residues. The first concerns the use of light to decompose transported pesticides in water and hence its application to irrigation waters and soils. The second concerns the decontamination of field soils containing DDT by flooding and inoculation with microorganisms.

II. Pesticide decontamination in water

A major source of environmental contamination is caused by the movement of materials from their site of application. Pesticides move primarily in the liquid or vapor phase. Pollution of water by organic compounds is undesirable, contamination by biologically active compounds is potentially dangerous. Two particular situations in which pesticides in water cause concern relate to the waste problem encountered in static or lagoon operations and to irrigation systems. The danger of the latter situation is best illustrated with the water-soluble, mobile herbicide picloram (4-amino-3,5,6-trichloropicolinic acid). Minute amounts of this potent herbicide irrigated on sensitive crops could have disastrous results. Concentrations as little as 10 p.p.b. in soils have a lethal effect on such sensitive crops as soybeans.

What methods are available for removing pesticides in water? The use of energetic radiation (ultraviolet or gamma ray) has been suggested (MARCUS *et al.* 1962) for fragmentation or destruction of organics in water. This method should be effective on a large number of pesticides especially in dilute aqueous solutions. Unfortunately, the technology has developed little beyond the experimental stage. Large scale ultraviolet and gamma irradiation techniques are in early technological stages and wider industrial application is the needed stimulus for further development.

We have determined the periods of exposure required to destroy the biological activity of a number of herbicide solutions in a small-scale ultraviolet irradiator. The method may be applicable as a pretreatment for waste waters or as a treatment for contaminated irrigation systems.

The reaction system is a borosilicate glass vessel and holds 250 ml. of the solution to be irradiated (Fig. 4). A quartz, water-cooled, double-walled tube is fitted into this well and is immersed in the solution. A 450 watt Hanovia lamp is suspended in the well. The quartz well transmits a large part of the available energy down to the shortest wavelengths emitted by the lamp.

Solutions of herbicides in water were irradiated (250 ml. at a time) for periods of 5, 10, and 15 minutes. Picloram, 2,4,5-T, bromacil, diphenamid, and 2,3,6-TBA were the herbicides used in the initial experiments. These compounds were chosen because their solubility and persistence are sufficiently high for them to be potential contaminants in irrigation water. Oats were used to bioassay picloram, 2,4,5-T, and diphenamid and cucumber was used for bromacil and 2,3,6-TBA. The treatments consisted of zero, 5, 10, or 15 minute irradiations of the solutions at 1, 5, or 10 p.p.m. concentrations and a control in which no herbicide was added. The time required to destroy the five herbicides is shown in Figure 5 (PLIMMER 1968).

A five-minute irradiation greatly reduced the phytotoxicity of picloram and 2,4,5-T at five and 10 p.p.m. and bromacil at one p.p.m. Diphenamid and 2,3,6-TBA required a 10-minute exposure. These initial results indicate that five minutes or less exposure to ultraviolet irradiation of solutions in the range of one p.pm. would significantly lower their phytotoxicity to plants. More pesticides, under conditions

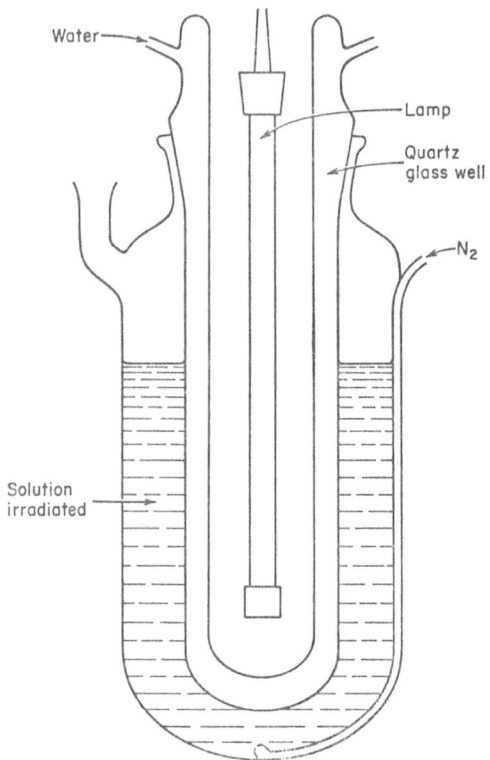

Fig. 4. Photochemical reaction vessel

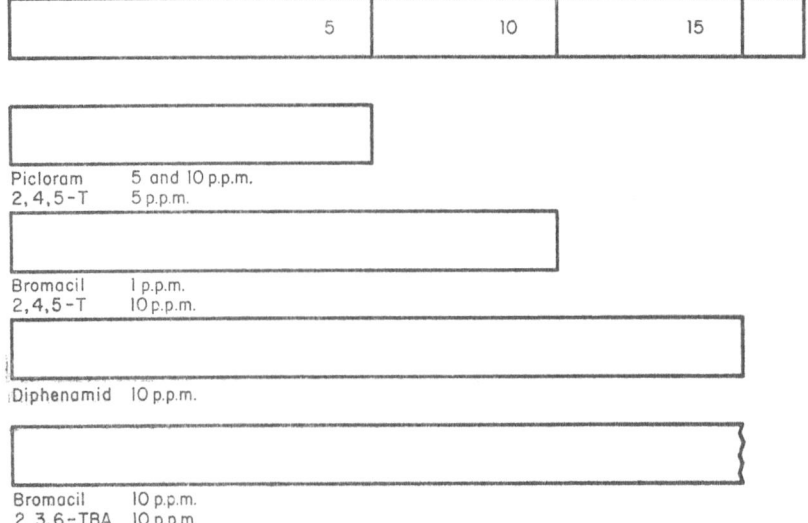

Fig. 5. Times required to destroy herbicidal activity by irradiation

approaching large volumes of water on flow systems need to be investigated before wide application to pesticide decontamination is attempted.

III. DDT decontamination

Turning our attention to field soils, one of the most serious residue problems occurs with organochlorine insecticides. As previously mentioned, DDT persists for several years in most agricultural soils. Complete removal of these residues may be impossible. However, lowering existing residues below some arbitrary threshold level may result in minimal plant residues. Our objective, then, was to find some agent in nature that could attack the DDT molecule. This agent did not necessarily have to cause complete destruction of DDT, but perhaps alter it to a more biodegradable or labile form. Obviously, organisms indigenous to most soils do not possess this agent. Intestinal microorganisms in the rat, however, are able to alter DDT extensively.

Whole cells or cell-free extracts of *Aerobacter aerogenes* catalyze the degradation of DDT *in vitro* to at least seven metabolites (WEDE-MEYER 1966), previously reported from rats given DDT orally (PETER-SON and ROBISON 1964). These reactions proceed by dechlorination, elimination, oxidation, and finally decarboxylation to yield dichloro-benzophenone. Therefore, it occurred to us soils inoculated with *A. aerogenes* may be capable of metabolizing residual DDT.

To test this hypothesis, three soil types were amended with zero, 5, 10, and 20 p.p.m. of DDT. The soils were Lakeland sandy loam, Hagerstown silty clay loam, and Sharkey clay. Four-hundred g. of soil were weighed into pots and DDT was applied in chloroform solution. Since metabolism of DDT by *A. aerogenes* appears to occur most rapidly in still cultures or under partially anaerobic conditions, two-thirds of the soils were flooded to simulate partial anaerobiosis. The water covered the soils to a depth of approximately one inch. One-third of the DDT-treated soils was maintained at field capacity, one-third was flooded, and one-third was flooded and inoculated with *A. aerogenes*.

Cells of *A. aerogenes* from slants obtained from the American Type Culture (ATC 13048) were mass cultured in three percent trypticase soy broth at 36° C. for eight hours. The cells were harvested by high-speed centrifugation, washed, and resuspended in the original volume of fresh broth solution. The cells were incubated for three days in still cultures, harvested again, washed, and concentrated 10-fold in a one percent yeast-extract solution. Aliquots (10 ml.) were added to the flooded soils and mixed into the surface layers. All soils were sampled at weekly intervals. Residual DDT and products were measured by electron-capture gas chromatography. Moist soil samples were extracted with a 3:1 mixture of hexane:isopropanol and injected on to a column of five percent SE 30 on 100/120 mesh DMCS-treated Chromo-sorb W. Column temperature was 210° C. with a flow rate of 120 ml./minute. Detector temperature was 215° C.

A total of 18 different parameters could be examined consider-ing there are possible three soils, three concentrations, and two treat-ments with *A. aerogenes.* Of primary interest is the effect of flood-ing with and without inoculation. Therefore, let us examine the disappearance of only DDT at the highest rates of application in the three soil types (Fig. 6). Two general trends are apparent. First DDT

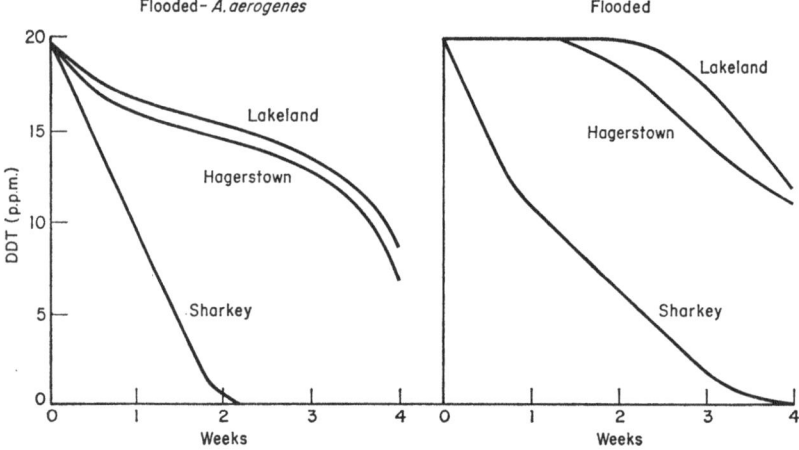

Fig. 6. DDT decomposition in three soils (see text)

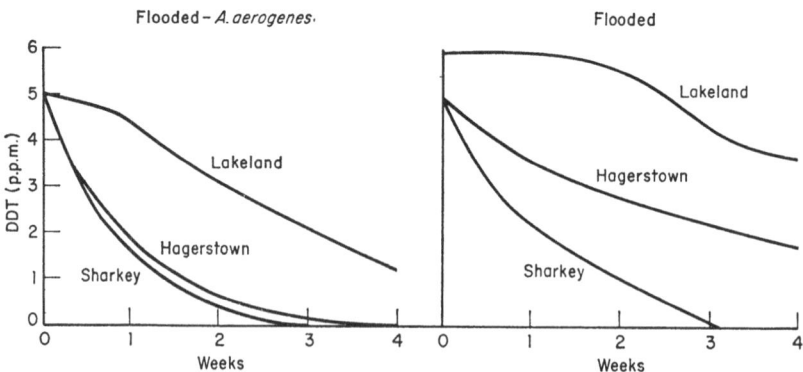

Fig. 7. DDT decomposition in three soils (see text)

disappeared more rapidly in the inoculated soils. Second, complete loss occurred in the Sharkey clay, while lesser amounts were lost from Lakeland and Hagerstown in both flooded series. DDD (TDE) was observed to occur in most soils, but its appearance did not parallel losses in DDT. In other words, there was a net loss of DDT-DDD in this system and no other products were detected by gas chromatography. Somewhat the same picture is encountered at five p.p.m. of DDT (Fig. 7), with complete loss in the Sharkey clay, more accelerated disappearance in the inoculated soils, and a general trend for total loss of DDT-DDD with time. Recovery values for DDT during the early sampling periods on Lakeland were high and explain the values obtained above five p.p.m. in the flooded soils.

Additional studies were initiated to determine the fate of the DDT in these systems. The experiment using Lakeland soil at five p.p.m. plus DDT-^{14}C was set up in a closed glass system for trapping $^{14}CO_2$. Nitrogen was bubbled through the system and any gaseous carbon dioxide was trapped in base. Samples of the carbon dioxide trapping solution were removed periodically and analyzed for radioactivity. Less than one percent of the total added activity was released in volatile components. This would be in general agreement with results previously reported (GUENZI and BEARD 1967).

Therefore, several other alternatives are available to explain the disappearance of the DDT from these soils. A polar metabolite could be present in the aqueous phase and not removed by the hexane:isopropanol extraction, or a metabolite is absorbed on some soil component and not recovered. It is also possible that a volatile metabolite is formed which escapes from the system with time. The presence of a polar metabolite can apparently be ruled out, since only 7 to 28 percent of the added DDT-^{14}C could be detected in the aqueous phase. Subsequent research with ^{14}C-DDT in a similar type of experiment indicated that up to 20 percent of the ^{14}C could not be extracted from flooded soils after a four-week incubation period. The radioactivity is apparently tightly bound to soil particles in some form not recoverable with hot hexane:acetone, ethyl acetate, or ethanol (KEARNEY and WOOLSON 1969). Therefore, we must conclude the loss is real and reproducible, although the mechanism is not fully understood.

IV. Conclusions

Radiation as a method for preventing the spread of pesticides in water systems deserves further consideration. Ultraviolet radiation for sterilization processes is in commercial use. High energy radiation has similar applications and its use in food processing has been studied. Ultraviolet radiation has been used for the complete removal of organic materials from sea water samples on a laboratory scale (ARMSTRONG et al. 1966). A pebble-bed type of reactor has been described, but not further developed, which appears suitable for continuous ultraviolet irradiation of solutions. A radioactive material is incorporated into impervious ceramic "pebbles" together with a suitable phosphor which emits ultraviolet radiation. A flow system is envisaged with the pesticide in dilute solution flowing over the ceramic pebbles. In addition to thermal methods of destruction, we suggest that radiation methods be further explored as a simple means of removing pesticides in situations where applicable.

Removal of pesticides from soils is a far more complex process, since the system is static and not conducive to flow-through operations. Several methods have been suggested for reducing residues. The use of calcium polyphosphate on the residual chloro-s-triazines has not been successful under field conditions (HARRIS et al. 1968). The use of absorbents (charcoal) for removing toxic materials in replanting certain nursery stock has been successful; extention of this technique to field conditions for atrazine and organochlorine residues has been attempted (LICHTENSTEIN et al. 1968). In these situations, however, the cure may be worse than the sickness, since a new and far less understood variable is now being introduced into the soil. Such may be the case with microbial decontamination of DDT by A. aerogenes. Many experimental variables would have to be studied before large-scale field studies would be justified.

In the final analysis any decontamination method would have to be economically feasible before it would be acceptable to the farmer. The most promising and yet still unexplored method for reducing soil pesticide residues lies at the molecular level. A thorough understanding of the electronic and steric factors that render a pesticide molecule susceptible to the natural biological degradation pathways is still in early developmental stages. This approach would appear to offer the most challenging chemical method for reducing residues on a continuing basis.

Table I. *Common and chemical names of pesticides mentioned in text*

Bromacil	5-bromo-3-sec-butyl-6-methyluracil
DDD	1,1-dichloro-2,2-bis(p-chlorophenyl)ethane
DDT	1,1,1-trichloro-2,2-bis(p-chlorophenyl)ethane
Diphenamid	N,N-dimethyl-2,2-diphenylacetamide
Picloram	4-amino-3,5,6-trichloropicolinic acid
2,4,5-T	2.4.5-trichlorophenoxyacetic acid
2,3,6-TBA	2,3,6-trichlorobenzoic acid

Summary

A limited number of methods are available for decontaminating soils and water. Complete destruction of organic herbicides in water may be effected by exposure to intensive high-energy radiation. Measurement of the rates of photodecomposition of picloram, diphenamid, bromacil, 2,3,6-TBA and 2,4,5-T by bioassay techniques is in progress. Experiments with model continuous flow cells indicate rapid destruction of organic dyes. No picloram could be detected in a solution of initial concentration of 1 p.p.m. after 30 minutes or less exposure to high intensity UV irradiation. Where complete removal is not feasible, reduction of existing residues below a level in soils where the significance of plant uptake becomes minimal may be desirable. Biological alteration of a persistent pesticide to a more degradable form is another method of reducing residues. DDT residues in three soils (Sharkey clay, Hagerstown silty clay loam, and Lakeland sandy loam at rates of 5, 10, and 20 p.p.m.) were reduced in flooded soils and flooded, enriched soils inoculated with *Aerobacter aerogenes*. Losses were most rapid in the inoculated Sharkey and Hagerstown soils receiving the lowest rate of DDT application during the first week. Parallel experiments conducted with ring labeled DDT-^{14}C showed no $^{14}CO_2$ evolved from the inoculated soils. Conventional chromatographic and radiometric techniques indicated a conversion of DDT to DDD and a reduction of the total DDT-DDD residue in soils with time.

Résumé [*]

Décontamination de pesticides dans les sols

On dispose d'un nombre limité de méthodes décontamination des sols et des eaux. On peut effectuer une destruction des herbicides organiques dans l'eau par exposition à des radiations intenses de haute énergie. La mesure des taux de photodécomposition du piclorame, du diphenamide, du bromacile, du 2,3,6-TBA et du 2,4,5-T par bio-essais est en progrès.

Des expériences à l'aide de cellules à courant continu indiquent une destruction rapide des colorants organiques. Aucune trace de piclorame n'a pu être décelée dans une solution qui en contenait initialement 1 p.p.m., après 30 minutes ou moins d'exposition à une irradiation UV de haute intensité. Dans les cas où une élimination complète n'est pas possible, une réduction des résidus présents en dessous d'une certaine limite peut être souhaitable pour les sols où l'importance de l'absorption par les plantes devient minime. La transformation biologique d'un pesticide persistant en une forme plus aisément décomposable est une autre méthode de réduction des résidus. Des résidus de DDT dans trois sols (argile de Sharkey, limon argileux alluvionnaire de Hagerstown et limon sableux de Lakeland aux concentrations de 5, 10 et 15 p.p.m.) ont été réduits dans des sols inondés et des sols inondés enrichis, inoculés avec *Aerobacter aerogenes*. Les pertes ont été plus rapides dans les sols de Sharkey inoculés et les sols de Hagerstown ayant reçu la plus faible concentration en DDT durant la première semaine. Des expériences parallèles avec du DDT marqué ^{14}C n'ont révélé aucun dégagement de $^{14}CO_2$ des sols inoculés. Les techniques chromatographiques et radiométriques conventionnelles ont indiqué une conversion du DDT en DDD et une réduction des résidus totaux de DDT-DDD dans les sols en fonction du temps.

---

[*] Traduit par S. DORMAL-VAN DEN BRUEL.

Zusammenfassung *

Dekontamination von Pestiziden im Boden

Eine begrenzte Zahl an Methoden steht zur Verfügung, um Böden und Wasser zu reinigen. Vollständige Zerstörung von organischen Herbiziden in Wasser kann durch Belichtung mit intensiver Strahlung von hoher Energie bewirkt werden. Messungen der Photoabbauraten von Picloram, Diphenamid, Bromacil, 2,3,6-TBA und 2,4,5-T durch Biotesttechniken sind im Fortschreiten begriffen. Experimente mit Modelldurchflusszellen zeigen schnelle Zerstörung von organischen Farben. In einer Lösung mit einer Anfangskonzentration von ein p.p.m. konnte nach 30 Minuten oder weniger Belichtung mit hoch intensiver ultravioletter Strahlung kein Picloram mehr nachgewiesen werden. Da, wo vollständige Entfernung nicht möglich ist, wird die Reduzierung von vorhandenen Rückständen in Böden unter eine Menge, wo die Bedeutung für die Pflanzenaufnahme minimal wird, wünschenswert. Biologische Veränderung eines persistenten Pestizids zu einer abbaufähigeren Form ist eine andere Methode, um Rückstände zu reduzieren. DDT Rückstände in Böden (Sharkey Ton, Hagerstown sandiger Ton-Lehm und Lakeland sandiger Lehm mit Raten von 5, 10 und 20 p.p.m.) wurden in überfluteten Böden reduziert und weggeschwemmt und angereicherte Böden mit *Aerobacter aerogenes* geimpft. Die Verluste waren am schnellsten in den Sharkey und Hagerstown Böden, welche während der ersten Woche die niedrigste Rate von DDT Behandlung erhalten hatten. Parallele Untersuchungen, welche mit ^{14}C-ring-markiertem DDT durchgeführt wurden, zeigten keine $^{14}CO_2$ Entwicklung in den beimpften Böden. Konventionelle chromatographische und radiometrische Techniken deuteten die Umwandlung von DDT zu DDD an und eine Reduktion des Gesamt-DDT-DDD-Rückstandes in Böden mit der Zeit.

References

Armstrong, F. A. J., P. M. Williams, and J. D. H. Strickland: Removal of organic matter from sea water by ultraviolet light. Nature 112, 481 (1966).

Guenzi, W. D., and W. E. Beard: Anaerobic biodegradation of DDT to DDD in soil. Science 156, 1116 (1967).

Harris, C. I., D. D. Kaufman, T. J. Sheets, R. G. Nash, and P. C. Kearney: Behavior and fate of *s*-triazines in soils. Adv. Pest Control Research 8, 1 (1968).

Kearney, P. C., and E. A. Woolson: Personal communication (1969).

——, R. G. Nash, and A. R. Isensee: Persistence of pesticide residues in soils. In M. W. Miller and G. C. Berg, ed.: Chemical fallout: Current research on persistent pesticides, Chapt. 3, pp. 54-67. Springfield, Ill.: Charles C Thomas (1969).

Lichtenstein, E. P., T. W. Fuhremann, and K. R. Schulz: Use of carbon to

---

* Übersetzt von A. Schumann.

reduce the uptake of insecticidal soil residues by crop plants. J. Agr. Food
Chem. **16**, 348 (1968).
MARCUS, R. J., J. A. KENT, and G. O. SCHENCK: Industrial photochemistry. Ind.
Eng. Chem. **54**, 20 (1962).
NASH, R. G.: Plant adsorption of dieldrin, DDT, and endrin from soils. Agron. J.
60, 217 (1968).
PETERSON, J. E., and W. H. ROBISON: Metabolic products of p,p'-DDT in the rat.
Toxicol. Applied Pharmacol. **6**, 321 (1964).
PLIMMER, J. R.: Unpublished data (1968).
WEDEMEYER, G.: Dechlorination of DDT by Aerobacter aerogenes. Science **152**,
647 (1966).

Interaction of diquat and paraquat with clay minerals and carbon in aqueous solutions *

By

SAMUEL D. FAUST ** AND ANTRA ZARINS ***

Contents

I. Introduction

In recent years, the demand for aquatic weed-free surface waters

---

* Paper of the Journal Series, New Jersey Agricultural Experiment Station, Rutgers, the State University of New Jersey, Department of Environmental Sciences, New Brunswick, N.J.

** Department of Environmental Sciences, Rutgers, The State University, New Brunswick, N.J.

*** New York Public School System, New York.

151

is on the increase. This is occasioned by greater and greater demands by the public for recreational and potable waters. One approach is to control aquatic weed growth in surface waters through the application of chemical herbicides. Several compounds and formulations are available of which diquat (1,1'-ethylene-2,2'-bipyridylium dibromide) and paraquat (1,1'-dimethyl-4,4'-bipyridylium dimethyl sulfate) are effective. These compounds are divalent cations and are quite soluble in water to the extent of 70 percent and greater. Furthermore, these compounds resist biological degradation in aquatic environments (HEMMETT 1968). Hence they may persist after an initial application for considerable periods of time and affect water quality for human consumption. The question of organic pesticide effects on water quality has been reviewed recently by FAUST and SUFFET (1966).

It is desirable, however, to prevent the transportation and distribution of organic contaminants via potable waters. This is especially true of organic pesticides where the effects of long-term consumption of trace quantities on human physiology are largely unknown. This paper discusses surface reactions of diquat and paraquat utilized in the conventional water treatment processes of carbon adsorption and chemical coagulation. Operational conditions of these processes were simulated as nearly as possible in the laboratory. Chemical oxidation of diquat and paraquat have been reported elsewhere by GOMAA and FAUST (1968). "Removal" is defined herein as the reduction in concentration of the two herbicides to an acceptable residual.

II. Literature review

HARRIS and WARREN (1964) offered one of the first reports on the adsorption of diquat on bentonite (calcium saturated), an organic soil (Houghton muck, 23 percent ash), an anion-exchange resin, and a cation-exchange resin. Since these studies were limited to 0.25 g. of adsorbent, adsorption capacities were not reported. For example, "Diquat was completely adsorbed by the bentonite and cation exchanger, and strongly adsorbed by the muck." On the other hand, the anion exchanger reached a capacity of approximately 7.0 μ moles/g. at an equilibrium concentration of 3 x 10^{-5} M diquat.

COATS et al. (1964) offered one of the first reports on the adsorption of paraquat on soil and clay minerals. A sandy loam soil (sand 66, silt 14, and clay 24 percent, and c.e.c. of 2.50 meq./100 g.) apparently adsorbed between 0.2 and 0.5 mg./g. (1.08 to 2.59 μ mole/g.) of paraquat. Kaolinite adsorbed between 2.5 and 3.0 mg./g. (13.4 to 16.1 μ mole/g.) whereas montmorillonite adsorbed between 75 and 85 mg./g. (40.3 to 45.7 μ mole/g.). No cation-exchange capacities nor cationic forms were given for the two clay minerals.

WEBER et al. (1965) and WEBER and SCOTT (1966) examined the influences of temperature and time on the adsorption of diquat and paraquat on kaolinite (c.e.c. of 0.051 meq./g., sodium form), mont-morillonite (c.e.c. of 0.847 meq./g., sodium form), and charcoal. At 10° and 55° C., diquat and paraquat were adsorbed in accord with the c.e.c. of the montmorillonite 427 μ mole (0.854 meg.)/g. Similar observations were made with kaolinite wherein adsorption for both herbicides was in an amount of 24 μ mole (0.048 meq.)/g. Darco G-60 charcoal (no adsorbent characteristics were given); adsorption within 48 hours was (a) 10° C., diquat = 60 μ mole/g., (b) 10° C., paraquat = 80 μ mole/g., (c) 55° C., diquat = 335 μ mole/g., and (d) 55° C., paraquat = 120 μ mole/g. These charcoal data were taken from the last point on the adsorption isotherms since WEBER et al. (1965) did not feel that equilibria were reached.

COATS et al. (1966) repeated their aforementioned studies for diquat. These quantities were adsorbed: (a) sandy loam soil, 0.3 mg./g. (1.64 μ mole/g.), (b) kaolinite, 2.0 mg./g. (10.9 μ mole/g.), and (c) bentonite, 80 to 100 mg./g. (43.5 to 54.4 μ mole/g.) "before significant quantities of diquat could be detected in the supernatant."

KNIGHT and TOMLINSON (1967) examined the interaction of para-quat with eight mineral soils at 20° ± 2° C. with 24 hours of contact. Langmuir adsorption isotherms were described where saturation up-take was less than the normal c.e.c. Typical quantities of adsorbed paraquat (meq./100 g.) were (a) sandy loam (c.e.c. = 2.88 meq./100 g.), 2.02, (b) loam (c.e.c. = 13.8, 9.8, and (c) clay (c.e.c. = 33.7), 25.7. Treatment with hydrogen peroxide eliminated most organic matter in the soils with a subsequent decrease in the c.e.c. and capacity to adsorb paraquat. Montmorillonite (c.e.c.=80) and kaolinite (c.e.c.=5.2) were equilibrated also with paraquat solutions: 78.0 and 4.0 meq./100 g., respectively. The Strong Adsorption Capacity ("a region of the adsorp-tion isotherm in which no paraquat could be detected in solution") of the eight soils was associated with the clay mineral fraction.

TUCKER et al. (1967) investigated the bonding of diquat and para-quat to five soil types: loam (c.e.c.=33.1 meq./100 g.), muck (112.7), sand (1.4), sandy loam (3.0), and silt loam (16.2). This bonding was arbitrarily defined as: "Loosely" bound–eluted with saturated am-monium chloride, and "tightly" bound–extracted by refluxing with 18N sulfuric acid. An "unbound" herbicide was eluted with water. In general terms, the sum of "loosely" and "tightly" bound paraquat (mg./g. dry soil) increased with the c.e.c. of the soil. On the other hand, this sum represented only 10 to 30 percent of the total c.e.c. for which no explanation was offered.

WEBER and COBLE (1968) researched the microbial decomposition of C^{14}-diquat adsorbed on montmorillonite and kaolinite. Soil micro-organisms were adapted to diquat in an aerobic system on an organic nutrient medium. $C^{14}O_2$ was inhibited completely in the montmorillon-ite systems when the appropriate quantity of clay was present, i.e., when the c.e.c. was satisfied with diquat. On the other hand, adsorption of the C^{14}-diquat by kaolinite did not inhibit $C^{14}O_2$ evolution. Presumably the difference between the two clay systems can be ascribed to adsorption of diquat into the crystalline lattice of montmorillonite where it is not available to microorganisms.

Weber et al. (1968) re-examined the question of adsorption and desorption of diquat and paraquat on and from Darco G-60 charcoal. The Freundlich equation described these systems. At equilibrium concentrations of $5 \times 10^{-6}M$, the amounts of diquat and paraquat adsorbed were approximately 5.0 and 7.0 μ mole/g. of charcoal, respectively. The desorption isotherms (deionized water extractions) essentially fell on the adsorption isotherms which suggests equilibria were attained in each system.

A summary of the adsorption capacities is offered in Table I. Data from this study are included also.

Table I. *Summary of adsorption capacities of diquat and paraquat on clay minerals and carbon*

Amount adsorbed (μ mole/g.)			Reference
Kaolinite	Montmorillonite	Carbon	
Diquat			
24	427 [a]	60 [b]	Weber et al. (1965)
10.9	43.5-54.4	—	Coats et al. (1966)
—	—	5.0 [c]	Weber et al. (1968 b)
—	386	93 [d]	This study
Paraquat			
13.4-16.1	40.3-45.7	—	Coats et al. (1965)
24	427 [a]	80 [b]	Weber et al. (1965)
—	—	7.0 [c]	Weber et al. (1968 b)
20	390	—	Knight & Tomlinson (1967)
—	386	140 [d]	This study

[a] 10° and 55° C., 30 minutes.
[b] 10° C., 48 hours.
[c] 25° C., one hour.
[d] 20° C., 30 minutes.

III. Materials

Diquat dibromide monohydrate — analytical standard grade, *Chevron Chemical Company*, Richmond, California; 51 percent active cation by weight. Appropriate stock solutions were prepared by direct weighing, dissolution, and dilution.

Paraquat dichloride — analytical standard, *Chevron Chemical Company*, Richmond, California; 72 percent active cation by weight. Appropriate stock solutions were prepared by direct weighing, dissolution, and dilution.

Aqua Nuchar A, a grade of activated carbon commonly used in water treatment plants. The specific surface area is 550 to 650 m./^{2}g. and 95 percent, by weight, passed through a 325-mesh sieve (44 μ).

Bentonite, 325 mesh, from *E. H. Sargent & Company*, Springfield, N.J. The cation-exchange capacity was 86.0 meq./100 g. as sodium hydroxide and was found in the calcium form. The specific surface area was assumed in accord with DIAMOND and KINTER (1958) as 600 to 800 m.^{2}/g.

Bentonite, H form, Department of Soils and Crops, Rutgers, the State University, New Brunswick, N.J.

Any other inorganic or organic chemicals were of reagent or analytical grade.

IV. Experimental methods

a) Analytical

The ultraviolet technique of FAUST and HUNTER (1965) was utilized for diquat and paraquat where the wavelengths of maximum absorption are 306 and 256 mμ, respectively. The molar absorptivities are 18,760 and 19,790 1 moles^{-1}cm.$^-$1, respectively, against a distilled water blank on a Beckman DB recording spectrophotometer.

b) Carbon adsorption isotherms

Five, one-l. portions of one mg./l. (approximately $5.4 \times 10^{-6}M$) diquat or paraquat in aged tap water, were placed on a multiple, continuous, and constant temperature shaker. Activated carbon was added to these portions from a 1.0 mg./ml. stock suspension to final dosages of zero, 10, 20, 30, and 40 mg./l. After contact periods of 30 or 60 minutes, the carbon was removed by #42 filter paper. A control consisted of aged tap water with no herbicide and the highest carbon dosage which also served as blank in the spectrophotometer. These isotherms were repeated at temperatures of 10°, 20°, and 40° C. at initial pH values of 7.2 and 9.0.

c) Chemical coagulation

Conditions under which hydrous oxides of aluminum form a negatively charged floc in aged tap water are initial pH = 8.5 to 9.0, total alkalinity = 50 mg./l. as $CaCO_3$ (by addition of 20 mg./l. of calcium hydroxide), and aluminum sulfate concentrations *circa* 50 mg./l. These conditions were determined by appropriate experiments.

In the herbicide systems, six 500-ml. portions were placed under a multiple stirring machine to which various quantities of aluminum sulfate were added for concentrations of zero, 20, 40, 60, and 80 mg./l.; the pH of each solution was adjusted to the range of 8.5 to 9.0 by addition of 20 mg./l. of calcium hydroxide. These systems were stirred rapidly for one minute to induce floc formation then stirred slowly for 20 minutes. Floc was allowed to settle for 30 minutes. Residual diquat and paraquat concentrations were then determined in the supernatant.

d) Clay mineral adsorption isotherms

Six one-l. portions of the appropriate herbicide concentration in aged tap water were placed under a multiple stirring machine. Calcium hydroxide (20 mg./l.) was added for adjustments of total alkalinity and the pH value into the 8.5 to 9.0 range. Subsequently, the clay minerals were added in the following quantities: (a) H-bentonite: zero, 20, 30, 50, and 100 mg. and (b) calcium-bentonite, zero, 10, 25, 50, and 100 mg. These systems were stirred slowly then for 10 minutes after which 50 mg./l. of aluminum sulfate was added. After a flocculation time of 15 minutes, the floc was allowed to settle. Any residual, suspended clay mineral and herbicide were measured in the supernatant.

e) Turbidity measurements

In order to evaluate coagulation of the clay mineral systems cited above, a turbidity measurement was utilized. A standard percent transmission curve was prepared from the turbidity imparted to aqueous solutions by zero, five, 10, 20, 30, and 50 mg./l. of the clay minerals. Transmittance was measured then against a distilled water blank on a Fisher electrophotometer with a blue (425 mμ) filter. A linear plot of log percent transmittance versus clay mineral concentration (mg./l.) was obtained.

V. Experimental results

a) Adsorption model

The Freundlich model described diquat and paraquat adsorption on carbon surfaces:

$$\frac{X}{M} = KC^{1/N} \tag{1}$$

where X = the amount of herbicide adsorbed in mg./l., M = the amount of carbon required to adsorb X in mg./l., C = residual herbicide concentration in mg./l., and K and $1/N$ are empirical constants. These latter values are evaluated from the log form of equation (1) where $1/N$ is the slope and K is the X/M value when C = 1.0 mg./l.

Once the K and $1/N$ values are obtained from a given system, then the Freundlich equation is rearranged:

$$M = \frac{C_o - C}{K \, C^{1/N}} \qquad (2)$$

where C_o = an initial concentration and C = a desired residual concentration. Thus, the amount of carbon, M, can be calculated for any initial herbicide concentration to any predetermined and desired residual.

b) Carbon adsorption isotherms

1. Diquat. — Typical isotherms are seen in Figure 1 for diquat at

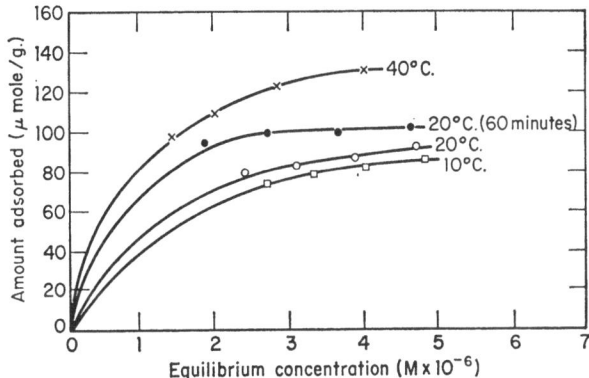

Fig. 1. Adsorption isotherms for diquat on active carbon: 30 minutes at pH 7.2

$10°$, $20°$, and $40°$ C., at a contact time of 30 minutes, and at an initial pH value of 7.2. Similar plots were observed for other experimental conditions noted in Table II, where the Freundlich constants are given. These constants were used, then, to calculate the carbon dosages required to reduce several initial concentrations of diquat to a residual of 0.1 mg./l. ($5.44 \times 10^{-7}M$) at contact times of 30 and 60 minutes. These are noted in Table III.

2. Paraquat. — Typical isotherms are seen in Figure 2 for paraquat at $10°$, $20°$, and $40°$ C., at a contact time of 30 minutes, and at an initial pH value of 7.2. Similar plots were observed for other experimental conditions noted in Table II where the Freundlich constants are given. These constants were used, then, to calculate the carbon dosages required to reduce several initial concentrations of paraquat to a residual of 0.1 mg./l. ($5.37 \times 10^{-7}M$) at contact times of 30 and 60 minutes. These are noted in Table III.

3. Effect of contact time. — The lower carbon dosages, Table III, suggest a considerable increase in diquat adsorption and a slight in-

Table II. *Freundlich constants for diquat and paraquat carbon systems* [a]

pH	Temp. (°C.)	30-Minute contact		60-Minute contact	
		K	1/N	K	1/N
		Diquat			
7.2	10	0.0164	0.26	0.0180	0.03
7.2	20	0.0174	0.20	0.0194	0.09
7.2	40	0.0262	0.25	—	—
9.0	20	0.0153	0.13	—	—
		Paraquat			
7.2	10	0.0258	0.10	0.0273	0.09
7.2	20	0.0253	0.09	0.0256	0.07
7.2	40	0.0308	0.16	—	
9.0	20	0.0268	0.09	—	

[a] Average of five determinations for each value.

Table III. *Effect of contact time on amounts of carbon required to reduce diquat and paraquat concentrations in water* [a]

Initial herbicide conc. (mg./l.)	Carbon dosage (mg./l.)			
	Diquat		Paraquat	
	30 Min.	60 Min.	30 Min.	60 Min.
1	82	57	44	41
2	173	121	92	87
3	264	184	140	133
5	445	311	238	225
7	627	438	335	317
10	900	629	480	456

[a] To residuals of 0.1 mg./l., at 20° C. and pH 7.2.

crease in paraquat adsorption when a 60-minute contact time is employed.

4. **Effect of temperature.** — That an increase in temperature increased the extent of diquat adsorption is seen in Figure 1 and Table IV. Less carbon was required to reduce equal concentrations of diquat to 0.1 mg./l. at 40° C. than at 20° and 10° C. On the other hand, temperature had a slight effect on the extent of paraquat adsorption at 40° over 10° and 20° C.

c) Chemical coagulation

Aluminum sulfate should form a negatively charged floc of

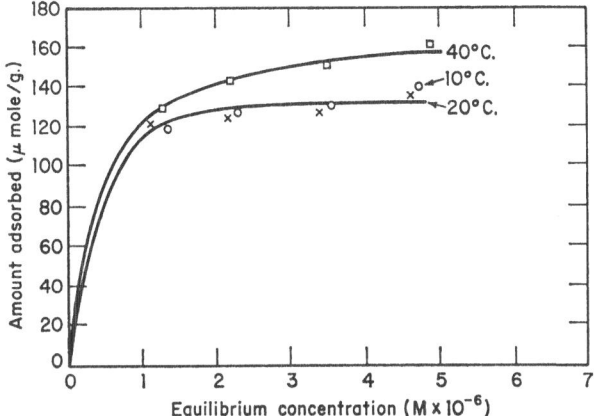

Fig. 2. Adsorption isotherms for paraquat on active carbon: 30 minutes at pH 7.2

Table IV. *Effect of temperature on quantities of carbon required to reduce diquat and paraquat concentrations in water* [a]

Initial herbicide conc. (mg./l.)	Carbon dosages (mg./l.) at		
	10° C.	20° C.	40° C.
	Diquat		
1	99.7	82	61
2	211	173	129
3	321	264	197
5	543	445	333
7	765	627	469
10	1100	900	674
	Paraquat		
1	44	44	42
2	93	92	89
3	141	140	136
5	239	238	230
7	337	335	323
10	484	480	465

[a] To a residual of 0.1 mg./l. in a contact time of 30 minutes at pH 7.2.

$Al(OH)_{3(s)}$ in accord with the general equation:

$$Al_2(SO_4)_3 + 3\,Ca(HCO_3)_2 \longrightarrow 2Al(OH)_{3(s)} + 3\,CaSO_4 + 6\,CO_2 \quad (3)$$

An attempt was made to adsorb the positively charged cations of diquat and paraquat on these negative flocs. That this was unsuccessful is seen in Table V where less than 10 percent of either initial concentration of diquat or paraquat was removed from the aqueous phase.

Table V. *Effect of chemical coagulation on the removal of diquat and paraquat from water* [a]

Aluminum sulfate conc. (mg./l.)	Initial herbicide conc. (mg./l.)	Conc. herbicide removed (mg./l.)
	Diquat	
50	0.1-10.0	0.08 [b]
60	1.06	0.07 [c]
80	1.06	0.07 [c]
100	1.06	0.08 [c]
	Paraquat	
50	0.1-10.0	0.12 [d]
60	1.06	1.10
80	1.06	0.09
100	1.09	0.09

[a] Twenty minute contact time.
[b] Average of 17 trials.
[c] Average of three trials.
[d] Average of 14 trials.

d) Clay mineral adsorption isotherms and chemical coagulation

That diquat and paraquat are adsorbed strongly and in accord with the cation-exchange capacity of clay minerals is seen in the literature review. This phenomenon can be utilized through the addition of a high exchange and specific surface area clay mineral (bentonite) to waters containing these herbicides. Chemical coagulation, then, removes any suspended bentonite. There is also the possibility that

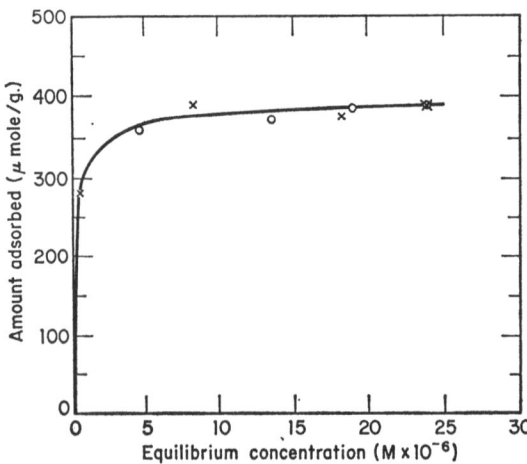

Fig. 3. Adsorption isotherm for diquat and paraquat on calcium-bentonite: O = diquat, × = paraquat

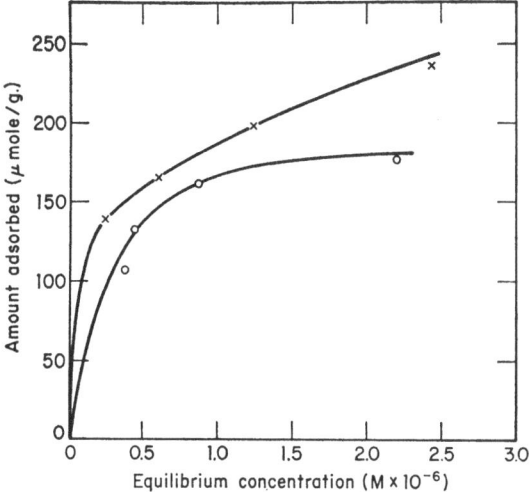

Fig. 4. Adsorption isotherms for diquat and paraquat on H-bentonite: O = diquat,
X = paraquat.

surface waters may transport natural silts and turbidity with these
herbicides adsorbed upon them. These particulates would be removed
subsequently in water treatment plants via chemical coagulation.

1. **Clay mineral adsorption isotherms.** — Typical isotherms are seen
in Figures 3 and 4 for the two bentonites from which the appropriate
Freundlich constants were derived. The quantities of bentonite required to reduce various concentrations of the herbicides to a residual
of 0.1 mg./l. are given in Table VI. Diquat and paraquat were ad-

Table VI. *Freundlich constants and amounts of the clay minerals required to
reduce diquat and paraquat concentration in water* [a]

Initial herbicide conc. (mg./l.)	H-bentonite (mg./l.)	Ca-bentonite (mg./l.)	H-bentonite (mg./l.)	Ca-bentonite (mg./l.)
1.0	36	15	29	16
2.0	75	33	60	34
3.0	115	50	92	52
5.0	194	84	157	89
7.0	273	118	218	125
10.0	391	170	317	180
K	0.044	0.066	0.052	0.067
1/N	0.24	0.054	0.22	0.084

[a] Ten minute contact time to a residual concentration of 0.1 mg./l.
[b] Room temperature.

sorbed to the same extent on the calcium-bentonite at an average of
386 μ mole/g. from the plateau of Figure 3. This approaches complete

saturation of the c.e.c. of the calcium-bentonite of 0.860 meq./g. These herbicides were adsorbed to a lesser extent on the H-bentonite as seen in Figure 4. This can be ascribed to a greater competitive effect of the H_3^+O ion for the exchange site (WAYMAN 1967).

2. Chemical coagulation — Verification of isotherms. — The data in Table VI represent predictions of the quantities of bentonite required to reduce various concentrations of diquat and paraquat to a residual of 0.1 mg./l. Subsequent experiments verified these predictions. Chemical coagulation effected the bentonite removal. That the Freundlich prediction was verified is seen in Table VII where diquat and paraquat

Table VII. *Diquat and paraquat removals from water via adsorption on clay minerals followed by chemical coagulation* [a]

Initial herbicide conc. (mg./l.)	Diquat			Paraquat		
	Ca-bentonite [b] (mg./l.)	Trans. (%)	Residual (mg./l.)	Ca-bentonite [b] (mg./l.)	Trans. (%)	Residual (mg./l.)
0.0	170	100	—	180	100	—
1.0	15.4	99	.04	16.3	97.2	.11
2.5	41	99.5	.02	—	—	—
5.0	84	97.5	.11	89	98.8	.05
7.5	127	99.5	.02	—	—	—
10.0	170	98.0	.09	180	99.8	.01

[a] $Al_2(SO_4)_3$ dosage = 50 mg./l., $Ca(OH)_2$ dosage = 20 mg./l, pH = 7.5.
[b] Initial contact time with clay mineral = 10 minutes followed by a 15 minute coagulation period.

residuals of 0.11 mg./l. and less were obtained. Transmittance values of 97.5 to 99.0 percent indicated that chemical coagulation effected virtually complete removal of the suspended bentonite.

3. Effect of contact time. — That a contact time between the clay mineral and the herbicide beyond ten minutes does not appreciably reduce residuals is seen in Figure 5. The uptake of diquat and paraquat occurs very rapidly with only a slight increase beyond ten minutes.

e) Adsorbent characteristics

The quantities of activated carbon, H-bentonite, and calcium-bentonite necessary to reduce various initial herbicide concentrations to 0.1 mg./l. are compared in Figures 6 and 7. Activated carbon is the least effective for diquat and paraquat when compared with the bentonites. Paraquat is more readily adsorbed on carbon and H-bentonite than is diquat. Adsorption on calcium-bentonite, however, is approximately the same for diquat and paraquat and it is the most effective of the three surfaces examined in this study.

VI. Discussion

Soluble organic contaminants are extremely difficult to remove from water. At potable water treatment plants the general approach to

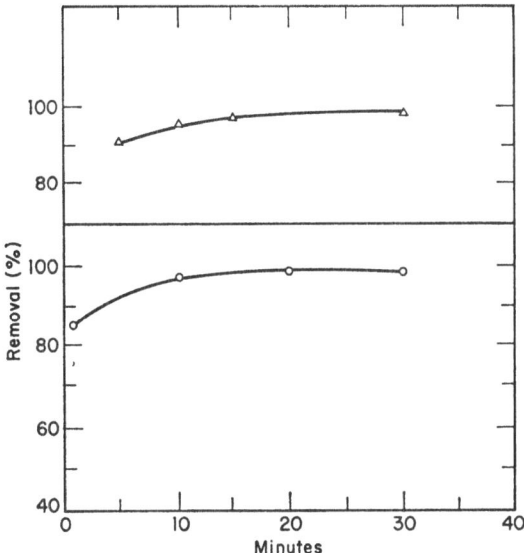

Fig. 5. Effect of contact time on adsorption of one mg./l. quantities of diquat and paraquat on H-bentonite (50 mg./l.): △ = diquat, ○ = paraquat

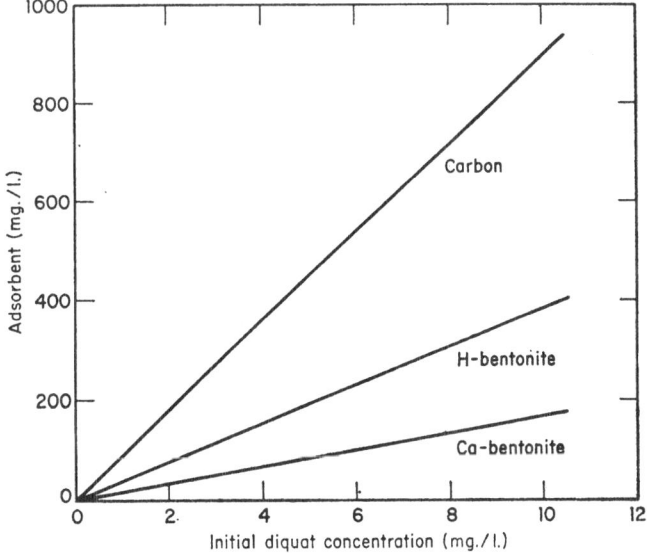

Fig. 6. Adsorbent dosages for removal of diquat from aqueous solutions

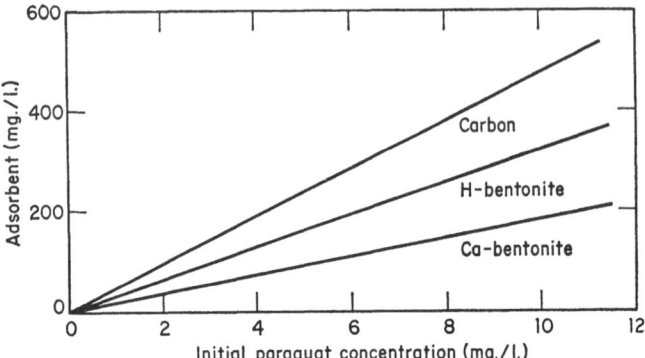

Fig. 7. Adsorbent dosages for removal of paraquat from aqueous solutions

this problem is to employ an adsorption reaction or a chemical oxidation reaction. In addition, the kinetics of these reactions must be
extremely favorable since the detention times of the plant reactors are
in the order of 15 to 60 minutes. Consequently, physical or chemical
reactions employed in the removal (*i.e.*, reduction in concentration) of
organic contaminants must go to completion in a short period. If, on
the other hand, the reaction kinetics are not favorable, compensation
must be made which is usually in the form of applying excess reactants
(carbon, oxidant, etc.). In other words, the reactions are forced to
completion via the law of mass action. Temperature is another reaction
variable that must be considered. Since reaction kinetics are usually
slower at lower temperatures, excess reactants are employed again to
force the reaction to completion. This, then, is the context under which
this study was conducted, namely, an evaluation in the laboratory of
adsorption reactions employed under treatment plant conditions of
detention time and temperature.

Carbon adsorption appears feasible for diquat and paraquat removal from aqueous systems within contact times of 30 and 60 minutes.
For example, carbon dosages of 82 and 44 mg./l. are required to
reduce an initial concentration of one mg./l. of diquat and paraquat,
respectively, to a residual of 0.1 mg./l. at 20° C. and at a contact time
of 30 minutes. These quantities of carbon are within normal operation
at a water treatment plant. Some operational advantage is gained in
the case of diquat through an increase in reaction time to 60 minutes
where the carbon dosage is less (57 mg./l.). A decrease in temperature from 20° to 10° C. does not appreciably affect diquat adsorption
whereas an increase to 40° C. lowers the carbon dosage (30 minutes)
to 61 mg./l. Temperature appeared to have little or no effect on
paraquat adsorption. A temperature of 40° C. has little practical
application at a water treatment plant but the data indicate that adsorption kinetics for diquat are more favorable at warmer tempera-

tures. Furthermore, paraquat appears to be adsorbed more easily than diquat as suggested by the lower carbon dosages. Apparently paraquat has a lower activation energy of adsorption than diquat since temperature had no effect. Conversely, diquat must have a higher activation energy than paraquat as evidenced by the higher carbon dosages, the effect of temperature, and the shape of the isotherms in Figure 1. In addition, diquat required a contact time of 60 minutes to reach equilibrium (isotherm in Figure 1).

Removal of soluble organic constituents from water is extremely difficult with negatively charged flocs of hydrous aluminum oxides. These flocs are usually employed for the destabilization of colloidal systems of inorganic and organic nature. That $Al(OH)_{3(s)}$ was unable to adsorb diquat and paraquat from aqueous solutions was seen in Table V. Consequently, the chemical coagulation processes at water treatment plants would not be expected to remove these herbicides from water. Apparently the negatively charged sites on the surface of $Al(OH)_{3(s)}$ are too weak to attract the cations of diquat and paraquat.

On the other hand, $Al(OH)_{3(s)}$ is employed for destabilization of colloidal systems. The latter are encountered quite frequently in natural waters serving as supplies for municipalities and industries. Moreover these colloid systems are composed of inorganic silts, clay minerals, soils, etc., that have been eluted from fields, river muds, and lake bottoms. Subsequent destabilization of these systems by chemical coagulation is necessary in a kinetic sense, namely, that the coagulant must form a floc, react with the colloid, and settle within 30 to 60 minutes of detention time in a water treatment plant.

Adsorption of diquat and paraquat on clay mineral surfaces is significant in two ways: (a) these herbicides may be present on the silts eluted from lakes treated for aquatic weed control or (b) clay minerals may be added at a treatment plant if these herbicides are found in a soluble state. If the latter approach is employed, then the kinetic factor of plant operation must be considered. Herein, bentonite (calcium form) adsorbed diquat and paraquat to capacity (in accord with the c.e.c.) within 10 minutes. Subsequent coagulation with $Al(OH)_{3(s)}$ effected removal of the residual turbidity (*i.e.*, bentonite) within 15 minutes. This combined reaction time of 25 minutes is well within normal operating conditions of water treatment plants. Data from the H-bentonite systems may indicate some of the reactions occurring in acid waters and bottom muds.

The bentonites were employed in this study because of their high c.e.c., surface areas, and expanding lattice. Diquat and paraquat are adsorbed strongly on this clay mineral. On the other hand, if the natural silts are mostly kaolinites and illites with their lower c.e.c. and surface areas then these herbicides would be adsorbed to a lesser extent and with lesser strengths.

It may be pertinent also to compare this study with the data of others (Table I). WEBER *et al.* (1965) employed a contact time of 30 minutes and temperatures of 10° and 55° C. for adsorption of diquat and paraquat on montmorillonite (sodium form). These investigators obtained an adsorption capacity of 427 μ mole/g. for both herbicides. Herein a capacity of 386 μ mole/g. was observed for a 10-minute contact time at room temperature. The agreement is excellent. COATS *et al.* (1965 and 1966) reported adsorption capacities lower by a factor of 10. Their systems were obviously not in equilibrium. KNIGHT and TOMLINSON (1967) are in agreement with this study, and with WEBER *et al.* (1965), with an adsorption capacity of 390 μ mole/g. for paraquat on montmorillonite.

On the other hand, the carbon adsorption data of this study and of WEBER *et al.* (1965 and 1968) are not in agreement. Higher capacities are reported herein: 93 μ mole/g. *versus* 60 μ mole/g. for diquat and 140 μ mole/g. *versus* 80 μ mole/g. for paraquat. Two reasons may be cited for this disagreement: (a) WEBER *et al.* employed carbon dosages of 2,500 mg./l. in one study and 5,000 mg./l. in the other. This would present extremely large surface areas in relation to initial herbicide concentrations in order to reach equilibrium and adsorption capacity. (b) The structure and surface functional groups of active carbons vary considerably, according to SNOEYINK and WEBER (1967). Aqua Nuchar A, commonly used in water treatment plants, is manufactured to have a high adsorption capacity. WEBER *et al.* (1965 and 1968) employed Darco G-60 charcoal for which adsorbent characteristics were not reported. Differences in adsorption capacities for these two studies may well lie in differences of surface functional groups and structure. There is qualitative agreement between these two studies wherein paraquat was adsorbd to a higher extent than diquat.

VII. Other treatment methods

There are two other possibilities for the removal of diquat and paraquat from water, namely, chemical oxidation (GOMAA and FAUST 1968) and adsorption on an ion-exchange resin (WEBER *et al.* 1968). The oxidation of diquat and paraquat by potassium permanganate and by chlorine confirm to second-order kinetics for which the reactions are pH dependent. Highest rates of oxidation were observed at pH values greater than 8.0. Chlorine dioxide oxidations of diquat and paraquat were extremely rapid in the pH range of 8.0 to 10.15 where no residuals were observed after one minute of reaction time. Chlorine is routinely employed in water treatment plants for disinfection purposes whereas potassium permanganate and chlorine dioxide can be employed whenever the occasion demands. The kinetics of these reactions are well within normal operating conditions.

Ion-exchange resins are not employed on a routine basis at water treatment plants but can be installed if necessary. WEBER *et al.* (1968)

employed an amberlite cation-exchange resin in H- and sodium-forms. Adsorption capacities of 237 to 250 μ mole/g. were observed for diquat and paraquat on both resin forms to approximately the c.e.c. Desorption was accomplished with $1M$ sodium chloride. This study suggests that an ion-exchange process would be quite feasible at water treatment plants for removal of trace quantities of diquat and paraquat from surface waters.

Acknowledgements

This study was supported by a grant from the *Chevron Chemical Company*, Ortho Division, Richmond, California. The authors gratefully acknowledge this support. This paper was presented, in part, before the American Chemical Society, Division of Agricultural and Food Chemistry, Atlantic City, N.J., Sept. 9, 1968.

Summary

The demand for recreational waters and potable water reservoirs to be free from aquatic weeds is on the increase. Diquat and paraquat have shown considerable promise for control of vegetation in aquatic situations. It is desirable, however, to prevent the transportation and distribution of pesticides via potable waters. The effects on human physiology of long-term consumption of traces of organic pesticides are largely unknown. This laboratory study evaluated carbon adsorption and chemical coagulation for the removal of diquat and paraquat from waters. Operational conditions of these processes were simulated as nearly as possible.

The Freundlich model described diquat and paraquat adsorption on active carbon surfaces. Freundlich constants from the isotherms provided carbon dosages for the reduction of various herbicide concentrations to an arbitrary 0.1 mg./l. For example, 82 and 44 mg./l. of carbon are required to remove 0.9 mg./l. of diquat and paraquat, respectively, from water within 30 minutes at 20° C. and pH 7.2. Freundlich constants are also given for 10° and 40° C. at 30 minutes and for 10° and 20° C. at 60 minutes. Carbon adsorption capacities were 93 and 140 μ mole/g. at 20° C. and 30 minutes for diquat and paraquat, respectively.

The negative surface of $Al(OH)_{3(s)}$ did not adsorb diquat and paraquat significantly, and it was concluded that chemical coagulation would not remove these highly soluble cations from water. Consequently, bentonite (Ca and H forms) was added in the appropriate quantities to adsorb diquat and paraquat within 10 minutes whereupon chemical coagulation followed for removal of any suspended clay mineral. Freundlich constants from the isotherms allowed calculation of the quantities of bentonite needed for reduction of herbicide concentration. Diquat and paraquat were more easily adsorbed on

Ca-bentonite than on H-bentonite than on carbon; paraquat was more easily adsorbed than diquat.

The utilization of carbon adsorption and chemical coagulation preceeded by clay mineral adsorption appears feasible for the removal of diquat and paraquat at conventional water treatment plants. Furthermore, these reactions appear feasible within normal operational conditions of detention times, pH, and temperatures.

Résumé *

Interaction du diquat et du paraquat avec l'argile et le charbon en solution aqueuse

La demande d'eaux exemptes de végétation aquatique pour les bassins de plaisance et les réservoirs d'eau potable est en progression. Le diquat et la paraquat se sont montrés riches en promesses pour le contrôle de la végétation aquatique. Il est cependant désirable d'empêcher le transport et la distribution des pesticides par les eaux potables. Les effets sur la physiologie humaine de l'absorption à long terme de traces de pesticides organiques sont grandement inconnus. Cette étude de laboratoire a évalué l'adsorption sur charbon et la coagulation chimique comme moyen d'élimination du diquat et du paraquat des eaux. Les conditions du terrain ont été simulées d'aussi près que possible. L'adsorption du diquat et du paraquat sur le charbon actif suit la loi de Freundlich. Les constantes de Freundlich déduites des isothermes ont permis de déterminer les doses de charbon nécessaires pour réduire les concentrations de divers herbicides au taux arbitraire de 0,1 mg/l. Par exemple, 82 et 44 mg de charbon par litre sont nécessaires pour enlever 0,9 mg/l de diquat et de paraquat, respectivement, en 30 mn de l'eau à 20° C et au pH 7,2. Les constantes de Freundlich sont aussi données pour 10° et 40° C à 30 mn et pour 10° et 20° C à 60 mn. Les capacités adsorbantes du charbon étaient respectivement égales à 93 et 140 μ mole/g à 20° et 30 mn pour le diquat et le paraquat. La surface négative de $Al(OH)_{3(s)}$ n'adsorbait pratiquement pas le diquat et paraquat et il fut conclu que la coagulation chimique n'enlèverait pas de l'eau ces cations très solubles. En conséquence de la bentonite (forme Ca et H) fut ajoutée en quantités appropriées pour adsorber le diquat et le paraquat en 10 mn, après quoi suivait une coagulation chimique pour soustraire tout argile minérale en suspension. Les constantes de Freundlich, d'après les isothermes, ont permis le calcul des quantités de bentonite exigées pour réduire la concentration de l'herbicide. Le diquat et le paraquat étaient plus facilement adsorbés sur la bentonite, forme Ca, que sur la bentonite H ou sur le charbon, le paraquat était plus facilement ad-

---

* Traduit par R. MESTRES.

sorbé que le diquat. L'utilisation de l'adsorption sur charbon et de la coagulation chimique précédée par une adsorption sur argile apparaît réalisable pour la suppression du diquat et du paraquat dans les installations conventionnelles de traitement des eaux. De plus, ces réactions semblent réalisables dans les conditions normales de durées des traitements de pH et de températures.

Zusammenfassung *

Wechselwirkung von Diquat und Paraquat mit Tonmineralien und Kohle in wässrigen Lösungen

Die Forderung, dass Wasser für Erholungszwecke und Trinkwasserreservoirs frei von Wasserunkräutern sein sollten, ist im Ansteigen begriffen. Diquat und Paraquat haben bedeutende Aussicht für die Vegetationskontrolle in Wassergegenden gezeigt. Jedoch ist es wünschenswert, den Transport und die Verteilung von Pestiziden über Trinkwasser zu vermeiden. Die Wirkungen auf die menschliche Physiologie bei langfristigem Gebrauch von Spuren von organischen Pestiziden sind weitgehend unbekannt. Diese Laborstudie wertet Kohleabsorption und chemische Koagulierung für die Entfernung von Diquat und Paraquat in Gewässern aus. Die Operationsbedingungen dieser Prozesse wurden so gut wie möglich nachgeahmt.

Das Freundlich-Modell beschreibt Diquat und Paraquatabsorption an Aktivkohleoberflächen. Freundlich-Konstanten von den Isothermen lieferten die Kohlemengen für die Reduktion von verschiedenen Herbizidkonzentrationen zu einem willkürlichen Wert von 0.1 mg/Liter. Zum Beispiel, 82 und 44 mg Kohle/Liter sind nötig, um 0.9 mg/Liter Diquat bezw. Paraquat aus Wasser innerhalb von 30 Minuten bei 20° C und pH 7.2 zu entfernen. Freundlich-Konstanten sind auch angegeben für 10° und 40° C bei 30 Minuten und 10° und 20° C bei 60 Minuten. Die Kohleabsorptionskapazitäten waren 93 und 140 μMol/g für Diquat bezw. Paraquat bei 20° C und 30 Minuten.

Die negative Oberfläche von $Al(OH)_{3(s)}$ absorbierte Diquat und Paraquat nicht wesentlich, und es wurde daraus geschlossen, dass chemische Koagulation nicht diese sehr löslichen Kationen aus Wasser entfernen würde. Folglich wurde Bentonit (Ca und H-Formen) in geeigneten Mengen zugesetzt, um Diquat und Paraquat innerhalb von 10 Minuten zu absorbieren, danach folgte chemische Koagulation zur Entfernung von möglichem suspendierten Tonmineral. Freundlich-Konstanten von den Isothermen erlaubte die Kalkulation der Bentonitmenge, die zur Reduktion der Herbizidkonzentration nötig ist. Diquat und Paraquat wurden leichter auf Ca-Bentonit als auf H-Bentonit und Kohle absorbiert. Paraquat wurde leichter absorbiert als Diquat.

---

* Übersetzt von A. SCHUMANN.

Die Ausnutzung von Kohleabsorption und chemischer Koagulation, welcher eine Tonmineralabsorption vorausgeht, scheint günstig für die Entfernung von Diquat und Paraquat bei konventionellen Wasserbehandlungsanlagen. Ausserdem scheinen diese Reaktionen innerhalb normaler Bedingungen der Zurückhaltungszeiten, pH und Temperaturen möglich.

References

COATS, G. E., H. H. FUNDERBURK, J. M. LAWRENCE, and D. E. DAVIS: Persistence of diquat and paraquat in pools and ponds. Proc. S. Weed Conf. 17, 308 (1964).

—— —— —— —— Factors affecting persistence and inactivation of diquat and paraquat. Weed Research 6, 58 (1966).

DIAMOND, S., and E. B. KINTER: Surface areas of clay minerals as derived from measurements of glycerol retention. In: Clays and clay minerals. NAS-NRC Publ. No. 566, p. 334 (1958).

FAUST, S. D., and N. E. HUNTER: Chemical methods for the detection of aquatic herbicides. J. Amer. Water Works Assos. 57, 1028 (1965).

——, and I. H. SUFFET: Recovery, separation, and identification of organic pesticides from natural and potable waters. Residue Reviews 15, 44 (1966).

GOMAA, H. M., and S. D. FAUST: Kinetics of chemical oxidation of dipyridylium quaternary salts. J. Agr. Food Chem., In press (1969).

HARRIS, C. I., and G. F. WARREN: Adsorption and desorption of herbicides by soil. Weeds 12, 120 (1964).

HEMMETT, R.: Private communication (1968).

KNIGHT, B. A. G., and T. E. TOMLINSON: The interaction of paraquat with mineral soils. J. Soil Sci. 18, 233 (1967).

SNOEYINK, V. L., and W. J. WEBER, JR.: The surface chemistry of active carbon. Environ. Sci. Technol. 1, 228 (1967).

TUCKER, B. V., D. E. PACK, and J. N. OSPENSON: Adsorption of bipyridylium herbicides in soil. J. Agr. Food Chem. 15, 1005 (1967).

WAYMAN, C. H.: Adsorption on clay mineral surfaces. In FAUST and HUNTER (ed.): Principles and applications of water chemistry, p. 142. New York: Wiley (1967).

WEBER, J. B. and H. D. COBLE: Microbial decomposition of diquat adsorbed on montmorillonite and kaolinite clays. J. Agr. Food Chem. 16, 475 (1968).

——, and D. C. SCOTT: Availability of a cationic herbicide adsorbed on clay minerals to cucumber seedlings. Science 152, 1300 (1966).

——, P. W. PERRY, and R. P. UPCHURCH: The influence of temperature and time on the adsorption of paraquat, diquat, 2,4-D and prometone by clays, charcoal, and an anion-exchange resin. Proc. Soil Sci. Soc. Amer. 29, 678 (1965).

——, T. M. WARD, and S. B. WEED: Adsorption and desorption of diquat, paraquat, prometone, and 2,4-D by charcoal and exchange resins. Proc. Soil Sci. Soc. Amer. 32, 197 (1968).

Kinetics of hydrolysis of diazinon and diazoxon *

By

H. M. GOMAA,** I. H. SUFFET,** AND S. D. FAUST **

Contents

I. Introduction

The widespread use of organic insecticides [1] in agricultural pest control has raised considerable concern over their potential contamination of water resources. More recently, however, direct analysis by gas-liquid chromatography indicates that most surface waters and some ground waters contain trace amounts of organic insecticides. Consequently, these raw water supplies may have to be monitored in order to avoid adverse effects from their ultimate domestic use as

---

* Paper of the Journal Series, New Jersey Agricultural Experiment Station, Rutgers, the State University of New Jersey, Department of Environmental Sciences, New Brunswick, N.J.

** Department of Environmental Sciences and Bureau of Conservation and Environmental Science, Rutgers, the State University, New Brunswick, N.J.

[1] Pesticides mentioned in text are identified in Table VI.

drinking water. Two general classes of insecticides are in wide use, the organochlorine compounds and the organophosphorus compounds. The former has high residual activity and has been implicated, in many cases, as the etiological agent for fish mortalities and other examples of deterioration of wildlife activity (ROSEN and MIDDLETON 1959, SWENSON 1962, NICHOLSON et al. 1964, WEISS and GAKSTATTER 1965). On the other hand, previous hydrolysis studies suggest that organophosphorus insecticides, under specific environmental conditions, may have low residual life in aqueous solutions.

Many organophosphorus compounds yield toxic metabolites by oxidation and isomerization before hydrolysis and detoxification occur (O'BRIEN 1960). Consequently, the analyst is confronted with a complex problem when determining these residues in polluted waters. A precise knowledge of the hydrolysis rates of organophosphorus insecticides and their metabolites (i.e., oxons) is of interest since chemical factors may determine whether or not toxic residues will persist. This investigation evaluates the kinetics of hydrolysis of diazinon and its oxygen analog, diazoxon, determines the activation energy requirements, and confirms the identity of the major hydrolysis product(s).

II. Literature review

PECK (1948) reported that parathion at 25° C. and pH less than 10 hydrolyzes at a rate of 50 percent in 120 days. KETELAAR (1950) determined the half-lives of parathion and methyl parathion at 15° C. in 1N sodium hydroxide solution as 32 and 7.5 minutes, respectively. WILLIAMS (1951) observed less than one percent hydrolysis of parathion in distilled water at 25° C. at pH 5 to 6. MUHLMAN and SCHRADER (1957) comprehensively investigated the influences of temperature and pH on hydrolytic stability of 21 organophosphorus pesticides. The conclusion was reached that under acid conditions and temperatures between 10° and 20° C. some stability was shown. The rate of hydrolysis increases markedly as hydroxyl ion activity increases which, in turn, decreases the half-life or persistence. Also, paraoxon hydrolyzes to diethyl phosphoric acid and p-nitrophenol at a rate approximately twice that of parathion at 20° C. and at pH values less than 5.0, but parathion hydrolyzes slightly faster than paraoxon at pH values greater than 5.0 at 70° C. GYSIN and MARGOT (1958) studied the chemistry and toxicological properties of diazinon and diazoxon. Diazoxon was found considerably less stable to hydrolysis than diazinon. At pH of 7.0 the rate was about ten times as high. The anticholinesterase activity of diazoxon proved to be higher than that for diazinon. EL-REFAI and GIUFFRIDA (1965) investigated the dehydrochlorination of Dipterex in fortified river water samples at pH 8.5 and 37° C. where conversion to DDVP was 50 percent complete in 24 minutes.

WEISS and GAKSTATTER (1965) studied the decay of diazinon by storage of water solutions of different pH values. Aqueous solutions of diazinon with 3.5 percent acetone at concentrations of 100 mg./l., with pH adjustments to 6, 7, and 8, were stored for 500 days at laboratory temperatures of 18 to 25° C. Diazinon appeared to be somewhat less stable at pH 6.0 than in the neutral or alkaline solutions. Qualitatively, from the graph of disappearance of diazinon activity *versus* time, the half-lives were determined as 25, 45, and 40 days at pH 6, 7, and 8, respectively. The half-lives were decreased by lowering the concentration of diazinon from 100 to 10 mg./l. The persistence of C^{14} labeled diazinon was determined to be similar in four soils under laboratory conditions at 25° C. by GETZIN and ROSEFIELD (1966). One-half of the original application was lost in two to four weeks and less than eight percent remained after 20 weeks. MORTLAND and RAMAN (1967) studied the catalytic hydrolysis of some organic phosphate pesticides by Cu (II). In the case of diazinon (0.008 mmole), the rate follows a second-order reaction with a half-life of four hours at 20° C. Diazinon was one of the organophosphorus pesticides studied by RUZIKA *et al.* (1967) who determined hydrolysis rates in ethanol at pH 6.0 buffer solutions (20:80) at 70° C. The half-life for diazinon was 37.0 hours. KONRAD *et al.* (1967) studied the mechanism of the fate of diazinon in soil systems, soil-free aqueous systems, and aqueous microbial medium inoculated with an aqueous soil extract. In an aqueous system at pH 6.0, diazinon was quite stable, while at pH 2.0 hydrolysis was extremely rapid. Apparently, microbial degradation was not a factor contributing to breakdown of diazinon in soils, although organisms capable of metabolizing diazinon may have been present. Using ring- and chain-labeled forms of diazinon in soil-free aqueous systems at pH 4.0 and 1.2 and sampling the hydrolysate at appropriate intervals, location of the C^{14}-label could be determined during the course of hydrolysis. Benzene removed intact diazinon. Extraction with *n*-butanol differentiated possible hydrolysis products.

III. Experimental techniques

a) Reagents

1. Stock solutions of diazinon and diazoxon were prepared by direct weighing of a reference grade material (99 percent pure) with dissolution in acetone (pesticide quality grade).
2. Sodium hydroxide solution – approximately 10N.
3. Phosphoric acid solution – approximately 5N.
4. The pH and ionic strength (0.02M) of reaction solution were maintained by addition of buffer salts. Orthophosphate solutions were used to cover the pH range from 3 to 11 (CHRISTIAN and PURDY 1962).
5. Hexane – pesticide quality grade.

6. Chloroform — pesticide quality grade.
7. Double-distilled water from alkaline permanganate.

b) Procedure for kinetics studies

Hydrolysis reactions were conducted in a thermostatically controlled water bath (±0.2° C.) in diffused light. Into a one-l. volumetric flask an aliquot of the insecticide stock solution was pipetted and reduced to near dryness with a gentle stream of nitrogen. Double-distilled water (900 ml.) was added; the flask was placed on a shaker for at least one hour to insure complete dissolution of the insecticide. Into another flask, exactly 100 ml. of the orthophosphate buffer solution was added. Both flasks were placed in the water bath to reach temperature equilibrium, after which the buffer solution was mixed with the insecticide solution. A stop watch was started at the initial pouring. At intervals aliquots of the reaction solution were withdrawn and poured immediately into a 250-ml. separator to which previously was added a desired amount of acid (H_3PO_4) or alkali (NaOH), for pH adjustment to 7.0 to 7.4. This separator also contained a desired volume of solvent for extraction of the residual insecticide (1:1 solution-solvent for diazinon, and 4:1 solution-solvent for diazoxon). After five minutes of shaking the lower aqueous layer was removed. The solvent layer was passed through a two-inch column containing approximately five g. of granular anhydrous sodium sulfate into a 50-ml. volumetric flask. The separator and the column were washed thrice with two ml. of solvent; washings were added to the eluate. The final volume of the extract was adjusted to 50 ml.

c) Choice of extraction solvent

At pH 7.0 to 7.4 hexane, benzene, ethyl acetate, and ether are excellent solvents for recovery of diazinon and diazoxon from water (Faust and Suffet 1968). Hexane was used since only diazinon or diazoxon was recovered from the aqueous phase and not the hydrolysis products including 2-isopropyl-4-methyl-6-hydroxyprimidine (IMHP). Chloroform was used for IMHP after recovery of diazinon or diazoxon with hexane.

d) Gas chromatographic determinations

A Microtek MT-200 gas chromatograph equipped with a 10 mc Ni[63] electron-capture detector at 250°C. was employed. Operating and chromatographic conditions with 15 percent Reoplex-400 and acid-washed Chromosorb W, 80 to 100 mesh, have been described (Suffet et al. 1967). Retention times for diazinon, diazoxon, and IMHP are 1.00, 1.45, and 1.77, respectively. A standard curve of peak height *versus* concentration was used to calculate the insecticide concentration in each triplicated sample in the hydrolysis series.

IV. Experimental results

a) Kinetic studies

The rate of hydrolysis of either diazinon or diazoxon was found to pH dependent, *i.e.*, the reaction was catalyzed by either $[H_3{}^+O]$ or $[OH^-]$:

$$\tag{1}$$

According to the law of mass action, the velocity should be dependent on the concentration of the two reactants:

$$-\frac{dC_{Ins.}}{dt} = k\, C_{Ins.}\, C_{cat.} \tag{2}$$

where, $C_{Ins.}$ is the residual concentration of diazinon or diazoxon (mole/l.) at time t. $C_{cat.}$ is the concentration of catalyst, *i.e.*, either $H_3{}^+O$ or OH^- in solution (mole/l.).

As the amount of catalyst consumed in the course of the reaction is negligible, its concentration may be regarded as constant. The velocity equation then takes the form:

$$-\frac{dC_{Ins.}}{dt} = K_{ob.}\, C_{Ins.} \tag{3}$$

so that the rate is proportional to the concentration of the insecticide. Thus, the process should follow first-order kinetics, as verified by experimental observations.

Integration of equation (2) yields:

$$t\, K_{ob.} = 2.303\, \log \frac{C_{oIns.}}{C_{Ins.}} \tag{4}$$

or,

$$\log C_{Ins.} = \log C_{oIns.} - \frac{K_{ob.}}{2.303}\, t \tag{5}$$

where $C_{oIns.}$ is the initial concentration of diazinon or diazoxon at time t (mole/l.). $K_{ob.}$ is the observed rate constant (t^{-1}).

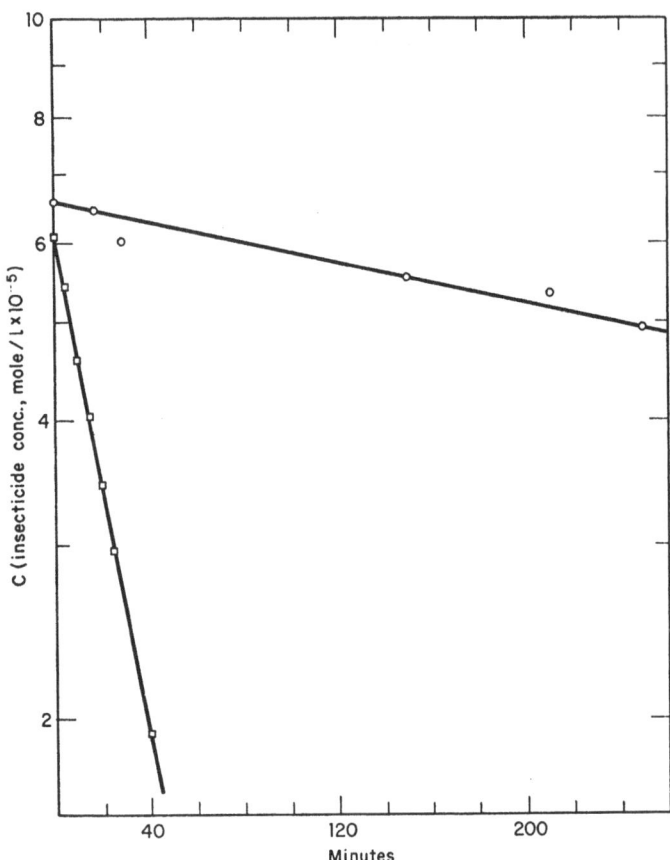

Fig. 1. Diazinon (○——○) and diazoxon (□——□) hydrolysis [see equation (5)]

Table I. *Diazinon and diazoxon hydrolysis at 20° C.*

pH	Diazinon [a]		Diazoxon [b]	
	$K_{ob.}$	$t^{1/2}$	$K_{ob.}$	$t^{1/2}$
3.1	9.8169×10^{-4} min.$^{-1}$	706.1 min.	3.0411×10^{-2} min.$^{-1}$	22.8 min.
5.0	9.3575×10^{-4} hr.$^{-1}$	740.7 hr.	2.2569×10^{-2} hr.$^{-1}$	30.7 hr.
7.4	1.5627×10^{-4} hr.$^{-1}$	4435.8 hr.	9.9953×10^{-4} hr.$^{-1}$	693.5 hr.
9.0	2.1244×10^{-4} hr.$^{-1}$	3263.0 hr.	1.5709×10^{-3} hr.$^{-1}$	441.2 hr.
10.4	4.7820×10^{-3} hr.$^{-1}$	144.9 hr.	6.8629×10^{-2} hr.$^{-1}$	10.1 hr.

[a] Initial concentration is $6.608 \times 10^{-5}M$, $I = 0.02M$.
[b] Initial concentration is $13.875 \times 10^{-5}M$, $I = 0.02M$.

Diazinon and diazoxon hydrolyses are plotted according to equation (5) in Figure 1. Good linear plots resulted from the first-order kinetic treatment; the second-order plots were not linear. Therefore, the rate can be represented by the half-life. Table I shows the hydrolysis rate constants and half-lives of diazinon and diazoxon at the different $H_3{}^+O$ concentrations at 20° C.

b) Activation energy requirements

We also determined the effect of temperature on the rate constants which, in turn, led to calculation of the activation energy of the hydrolysis reactions. The rate constant $K_{ob.}$ was evaluated at 10°, 20°, 40°, and 60° C. from the slope of a plot of log $C_{Ins.}$ as a function of time (equation 5) for each experimental run. The velocities of the reactions vary with temperature as shown in Tables II and III.

Table II. *Hydrolysis constants of diazinon* ^{a} *at different temperatures*

Temp. (°C.)	pH 3.1		pH 10.4	
	$K_{ob.} \times 10^{-4}$ (min.$^{-1}$)	$t_{1/2}$ (min.)	$K_{ob.} \times 10^{-3}$ (hr.$^{-1}$)	$t_{1/2}$ (hr.)
10	4.5211	1532.8	2.2114	313.4
20	9.8169	706.1	4.7820	144.9
40	39.3841	175.6	14.5732	47.5
60	147.9822	46.7	55.5834	12.4

^{a} Initial concentration is $6.608 \times 10^{-5}M$, $I = 0.02M$.

Table III. *Hydrolysis constants of diazoxon* ^{a} *at different temperatures*

Temp. (°C.)	pH 3.1		pH 10.4	
	$K_{ob.} \times 10^{-2}$ (min.$^{-1}$)	$t_{1/2}$ (min.)	$K_{ob.} \times 10^{-2}$ (hr.$^{-1}$)	$t_{1/2}$ (hr.)
10	1.4471	47.8	3.2101	21.6
20	3.0411	22.8	6.8629	10.1
40	11.3458	6.1	22.0046	3.1
60	40.1142	1.7	102.3752	0.7

^{a} Initial concentration is $13.875 \times 10^{-5}M$, $I = 0.02M$.

The Arrhenius equation was employed:

$$K_{ob.} = A_{ob.}\, e^{-E_{ob.}/RT} \tag{6}$$

where $K_{ob.}$ is the observed rate constant (t^{-1}),

$A_{ob.}$ is a constant known as the frequency factor,

$E_{ob.}$ is the observed activation energy of the reaction $(K \text{ cal.mole}^{-1})$,

T is the absolute temperature, and

R is a constant equal to 1.986 cal.mole^{-1}deg.$^{-1}$.

The logarithmic form of the Arrhenius equation is:

$$\log_{10} K_{ob.} = \log_{10} A_{ob.} - \frac{E_{ob.}}{2.303RT} \qquad (7)$$

It follows from equation (7) that a plot of the logarithm of the rate constant against the reciprocal of the absolute temperature should yield a straight line whose slope is $-\dfrac{E_{ob.}}{2.303R}$ or $-\dfrac{E_{ob.}}{4.574}$. The observed activation energy $E_{ob.}$ is calculated from this slope (Table IV).

Table IV. *Activation energy of the hydrolysis reactions*

Compound	pH	$E_{ob.}$ (Kcal.mole^{-1})
Diazinon	3.1	13.148
	10.4	14.286
Diazoxon	3.1	12.451
	10.4	13.095

That the data confirm the Arrhenius equation is seen in Figure 2. It appears that the rate of the reaction approximately doubled for each 10 degree increase in temperature. Slight deviations appear at the higher temperatures (40° and 60° C.). The calculated activation energy for diazoxon hydrolysis is generally smaller than that for diazinon under the same conditions of pH and ionic strength. The activation energy requirements were higher under alkaline conditions (pH 10.4). Since the hydrolysis rates of diazinon and diazoxon were slower under a neutral pH of 7.4, the activation energy requirements may be maximum here which decrease under acid or alkaline conditions.

c) Identification of major hydrolysis products

1. **Collection of gas chromatographic peaks.** — The gas-liquid chromatographic collection system of GIUFFRIDA (1965) with potassium bromide as an absorbent was modified for single column operation with a flame ionization detector (FAUST and SUFFET 1968). This technique collected sufficient amounts of diazinon, diazoxon, and IMHP for subsequent confirmatory identification steps.

2. **Confirmation by infrared and ultraviolet microspectrophotometry.** — In order positively to identify a GLC peak, it must be collected and

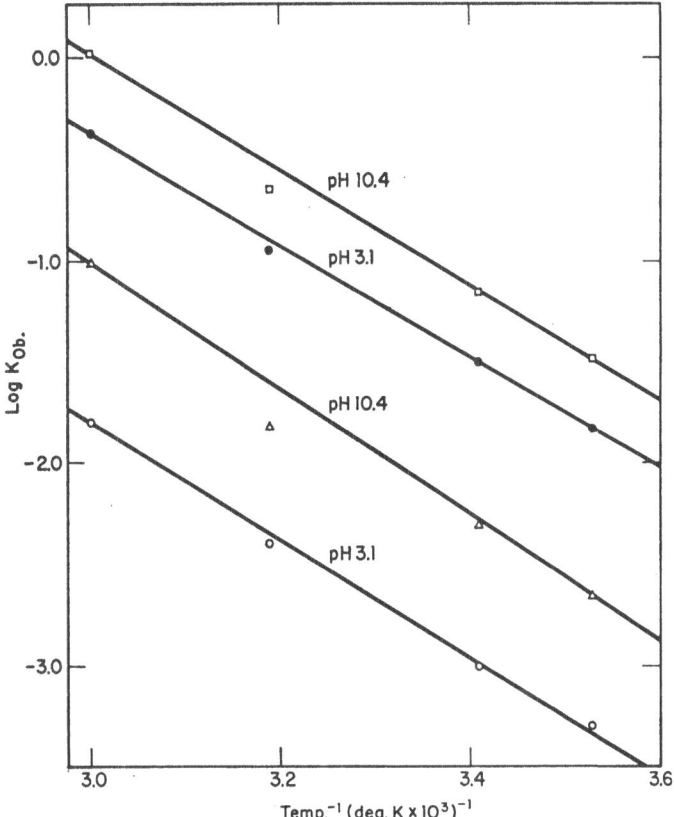

Fig. 2. Arrhenius—plot for diazinon ($\bigcirc$——$\bigcirc$ and $\triangle$——$\triangle$) and diazoxon ($\bullet$——$\bullet$ and $\square$——$\square$) hydrolysis

analyzed by ancillary techniques which should have the ability to identify molecular characteristics of pure compounds. In this study, confirmation of the parent and hydrolysis compounds was accomplished with infrared and ultraviolet microspectrophotometry (FAUST and SUFFET 1968).

a. **Infrared.** – An aliquot of the test solution, taken when at least 50 percent of the hydrolysis was complete, was extracted twice with hexane (1:1) to recover nearly 100 percent of the residual parent compound (diazinon or diazoxon). Three steps of 2:1 (solution-chloroform) extraction then were applied to recover nearly 95 percent of IMHP. Eluate hexane and chloroform were flash evaporated and then injected into the gas chromatograph provided with the collection system. The relative retention times of the parent compounds and hydrolysis product IMHP corresponded to those determined with standard solutions under the same conditions. The different peaks were collected. Figures 3, 4, and 5 show the infrared spectra of collected

H. M. GOMAA, I. H. SUFFET, AND S. D. FAUST

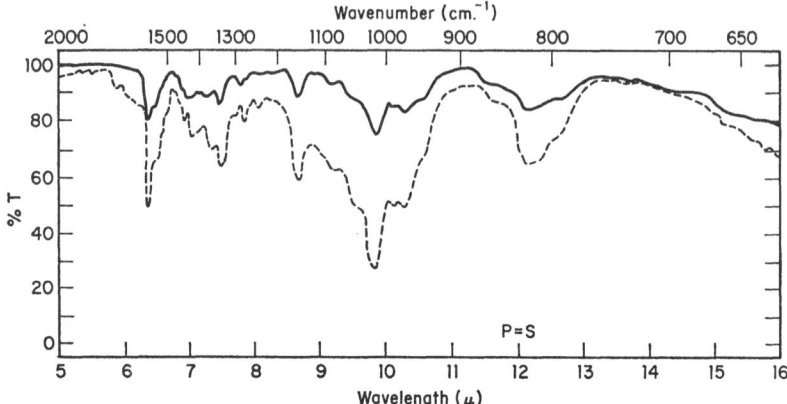

Fig. 3. Infrared spectra of standard (——) and collected (- - - - -) diazinon

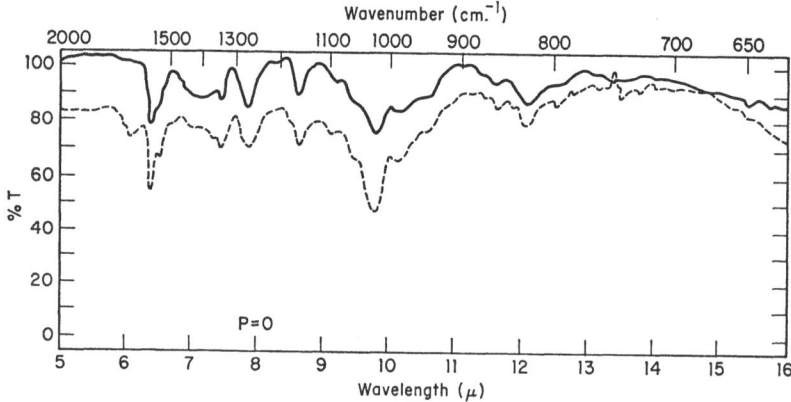

Fig. 4. Infrared spectra of standard (——) and collected (- - - - -) diazoxon

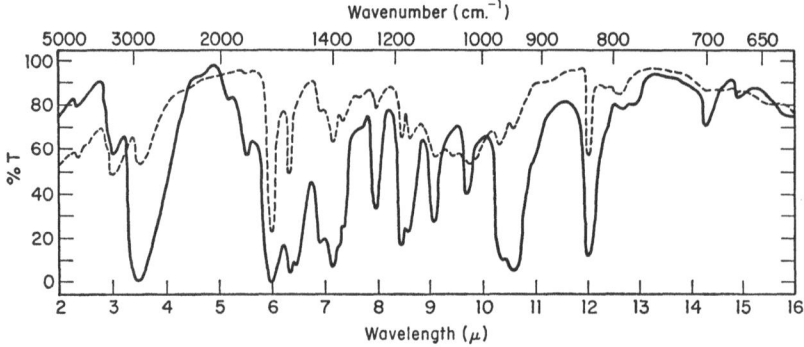

Fig. 5. Infrared spectra of standard (——) and collected (- - - - -) IMHP

GLC peaks versus standard spectra of diazinon, diazoxon, and IMHP, respectively.

β. **Ultraviolet.** — Ultraviolet spectroscopy has been utilized on micro-samples along with infrared analysis to supplement other confirmation techniques. GERSMANN and KETELAAR (1958) have discussed the ultraviolet spectra of phosphate esters. The ester part of the or-

$$O(S)$$
$$\|$$

ganophosphate pesticide esters $(RO)_2$-P-O(S)-X does not appear to absorb ultraviolet radiation, but appears only to effect the absorption by the X-group. In general, the ultraviolet spectrum of the hydrolysis product will be different from that of its parent and oxon due to an inherent chromophoric group (X) and the strong interaction of this group with the OH or SH group formed upon hydrolysis [X-OH(SH)] as seen in Figure 6. Figure 7 shows the standard and collected IMHP ultraviolet spectra.

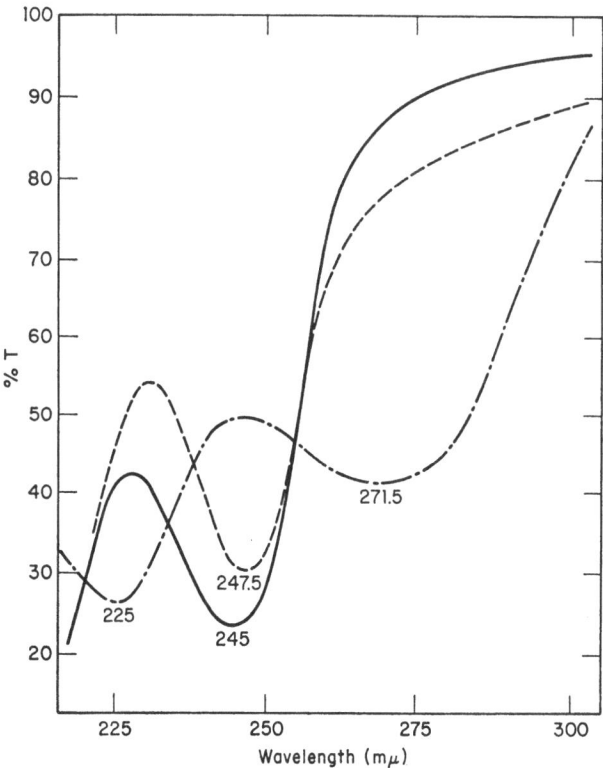

Fig. 6. Ultraviolet spectra of diazinon (———), diazoxon (- - - - -), and IMPH (– - – - –) in 95 percent ethyl alcohol, one-cm. cells

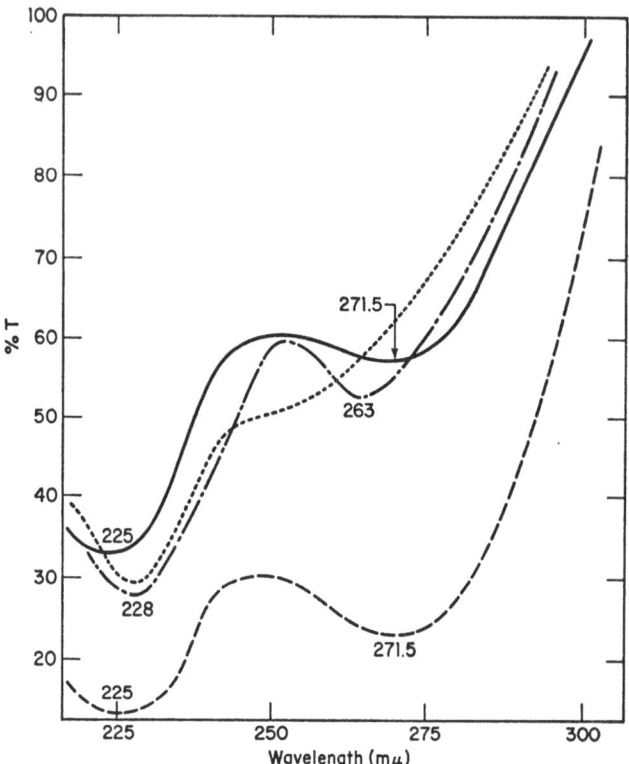

Fig. 7. Ultraviolet spectra of IMHP: ———— IMHP standard in 95 percent ethyl alcohol, – – – IMHP collected in 95 percent ethyl alcohol, ... IMHP standard in water at pH 3.1, and — · — · — IMHP standard in water at pH 10.4

d) Quantitation of hydrolysis product

Quantitation of hydrolysis product IMHP was accomplished as follows: The reaction solutions of diazinon and diazoxon at pH 3.1 and 10.4 were left in a water bath for five days at 50° C. to insure complete hydrolysis. An aliquot of the test solution was extracted with hexane which, after drying, was injected into the GLC electron-capture detector to verify complete disappearance of diazinon or diazoxon. Another aliquot of the test solution was measured at 228 mμ with the concentration of IMHP determined from an aqueous standard curve. The aqueous standard solutions were measured in the ultraviolet at two pH values (3.1 and 10.4) because of the shift of the spectra with $[H_3^+O]$ (Fig. 7). Hydroxypirimidines exist in two isomeric forms:

Ketone ⇌ Enol

Table V. *Ultraviolet determination of IMHP*

Compound	pH	IMHP (mg./l.)	
		Calc.	Exper.
Diazinon	3.1	10.000	10.012
	10.4	10.000	10.010
Diazoxon	3.1	21.115	20.552
	10.4	21.115	21.115

From Table V it is clear that diazinon and diazoxon are quantitatively hydrolyzed to IMHP and diethyl thiophosphoric or diethyl phosphoric acid according to the equation:

$$(8)$$

V. Discussion

An order of magnitude of persistence and/or appearance of hydrolysis products of diazinon and diazoxon in natural waters can be obtained from the kinetic data. In general, at 20° C., diazoxon hydrolysis proceeds much faster than diazinon under any comparable conditions. A large difference in rate of hydrolysis can be detected under acidic conditions, where diazoxon is hydrolyzed 30 times faster than diazinon. The differences in rates decrease as neutrality is approached, but increase again as the pH of the hydrolysis solution increases (diazoxon hydrolyzed seven times faster than diazinon at pH 7.4 and 14 times faster at pH 10.4).

FUKUTO (1957) stated that tertiary phosphate esters can be hydrolyzed readily to secondary esters. The bond which is broken depends upon structure of the compound and hydrolytic conditions. Studies of the mechanism of hydrolysis in O^{18}-rich water of a tertiary ester $(RO)_2$-$\overset{\overset{\displaystyle O}{\|}}{P}$-O-R', show that, in alkaline solution, the usually expected displacement of the OR' group occurs (Blumenthal and Herbert 1945, Chanley et al. 1952, Vernon 1957, Barnard et al. 1961). In acidic hydrolysis, a rupture of the O-R' apparently occurs as the first step. Usually the secondary esters are quite stable to alkaline hydrolysis, and further attack occurs only under fairly drastic conditions. In acid solutions, the hydrolysis of secondary esters proceeds rather rapidly, but generally at slower rates than shown by the tertiary esters (Plimmer and Burch 1929). The hydrolysis of simple organic phosphates in acid and in base, therefore, can be summarized (Fukuto 1957):

$$(RO)_2 - \overset{\overset{\displaystyle O}{\|}}{\underset{|}{P}} \!\!+\!\! O - R' \xrightarrow{\ OH^- \ } (RO)_2 - \overset{\overset{\displaystyle O}{\|}}{P} - O^- \ + \ R'OH \tag{9}$$

$$(RO)_2 - \overset{\overset{\displaystyle O}{\|}}{P} - O \!\!+\!\! R' \xrightarrow[H_2O]{\ H_3^+O \ } (RO)_2 - \overset{\overset{\displaystyle O}{\|}}{\underset{\displaystyle \downarrow}{P}} - OH \ + \ R'OH \tag{10}$$

$$\text{Primary ester} \longrightarrow H_3PO_4$$

Some organophosphorus compounds, however, do not follow this general rule of acidic and basic hydrolysis. Cleavage of the P-O or R'-O bonds depends upon the nature of the group R.

Hydrolysis of the RO- and P-O-R' bonds under alkaline conditions has been found in most cases to be catalyzed by OH⁻ and to be of the S_N2 type (Dostrovsky and Halmann 1953 and 1956, Heath 1956 a and b). That is to say: (a) it is a nucleophilic substitution reaction in which the OH⁻ substitutes for the group $(O\text{-}R')^-$

$$OH^- + \overset{\displaystyle RO}{\underset{\displaystyle RO}{>}}\!\!\overset{\overset{\displaystyle \delta-}{\overset{\displaystyle \cdots}{O(S)}}}{\underset{\displaystyle \delta+ \ \ \ }{P}}\!\!\overset{}{\underset{\displaystyle O-R'}{}} \ \rightleftharpoons \ \left[\overset{\displaystyle RO}{\underset{\displaystyle RO}{>}}\!\!\overset{\displaystyle O(S)}{\underset{\displaystyle \text{HO-\!-\!-\!P-\!-\!-\!-O-R'}}{}} \right] \ \rightleftharpoons \ \overset{\displaystyle RO}{\underset{\displaystyle RO}{>}}\!\!\overset{\displaystyle O(S)}{\underset{\displaystyle \text{HO-P}}{}} \ + \ (O-R') \tag{11}$$

(b) it proceeds by a bimolecular mechanism, and (c) no stable intermediate forms as the OH⁻ approaches the molecule and attacks the P-atoms. The latter has been made electrophilic by the inductive effect of the =O or =S. In some of the attacked molecules, $(O\text{-}R')^-$ is released and simultaneously OH⁻ combines with the P.

By replacing the P=O for P=S, a molecule of greater hydrolyzability should result. This is due to the greater electrophilic character of =O when compared with =S. The results herein support this (Tables I, II, and III). Also, this may explain why the activation energy requirement for diazinon is slightly higher than that for diazoxon under comparable conditions (Table IV). The calculated activation energies of diazinon are 13.148 and 14.286 Kcal mole⁻¹ at pH 3.1 and 10.4, respectively.

MORTLAND and RAMAN (1967) reported a value of 11 to 13 Kcal.mole^{-1} for the Cu(II)-catalyzed hydrolysis reaction in 1:1 water-methanol solution. The mechanism is suggested as a bidentate chelation through nitrogen in the ring structure and sulfur on the phosphate side chain in the diazinon molecule. This forms a closed cyclic resonance system; electron shifts in the proposed ring structure weaken the bonding of the side chain to pyrimidine thereby promoting hydrolysis.

Metabolism of diazinon in plants has been shown to involve hydrolysis of the phosphorus pyrimidyl ester bond and subsequent metabolism of the IMHP to carbon dioxide. Small amounts of diazoxon, also were detected in field grown crops (RALLS et al. 1966). However, KANSOUTH and HOPKINS (1968) reported that no diazoxon was detected in bean plant extracts or nutrient solution, indicating that oxidation is very minor or that oxon is hydrolyzed as rapidly as it is formed.

The pathway of diazinon hydrolysis and all possible intermediate compounds are illustrated in Figure 8 according to KONRAD et al.

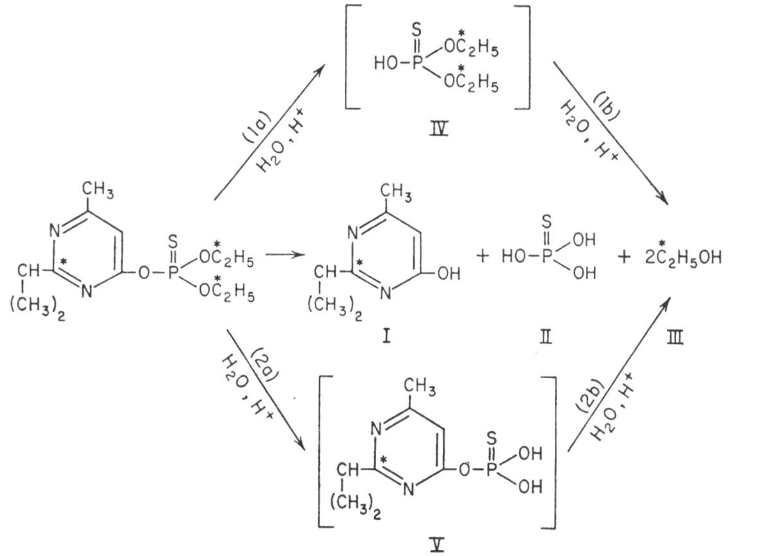

Fig. 8. Hypothesized hydrolysis of diazinon (after KONRAD et al. 1967); asterisks indicate location of labeled C^{14} atoms

(1967). A conclusion was reached, using chain-labeled diazinon at pH 4.0, that diethyl thiophosphoric acid was the predominant labeled product and diazinon hydrolysis proceeds via pathways 1a and 1b with the production of the unlabeled 2-isopropyl-4-methyl-6-hydroxypyrimidine I (IMHP).

Data in Table V confirm that pathways 1a and 1b in Figure 8 are the ultimate way of hydrolysis for diazinon and diazoxon under acidic and alkaline conditions. This has been confirmed by GLC, infrared, and ultraviolet techniques. However, any variations of pH may affect the specific bond that must be ruptured. Under acidic conditions a rupture of the O-R′ link (R′=the pyrimidyl radical) is expected, while in alkaline solution P-O cleavage usually predominates.

Since most organophosphorus pesticides hydrolyze, their persistence in natural waters may be controlled by chemical forces rather than by biological activities. In general, pH values of natural waters range from 5.5 to 8.5, whereas temperatures are less than 25° C., and diazinon and diazoxon will have a long residual life. According to O'BRIEN (1960), the conversion of P=S to P=O (thiophosphate esters to its oxygen analog) may be accomplished by chemical or enzymic oxidation which, in turn, increases their cholinesterase inhibition activity and hydrolyzability. Natural waters may provide the enzymic oxidation opportunity by aquatic insects, fish, and microorganisms. Chemical oxidation may be afforded by dissolved oxygen in natural waters or by chlorine and potassium permanganate that are used in conventional water treatment plants. Should oxygen analogs appear in natural or treated waters then their persistence and effect on potable water quality must be considered.

VI. Conclusions

Diazinon and diazoxon are hydrolyzed by first-order kinetics to IMHP in the pH range 3.1 to 10.4, at an ionic strength of 0.02M. Under natural water conditions diazinon and diazoxon are characterized by a long residual life. Under specific conditions (pH lower than 5.0 or higher than 9.0, and temperatures higher than 20° C.) hydrolysis proceeds within a relatively short period of time compared to the more persistent chlorinated hydrocarbons and phenoxy alkyl acids.

Acknowledgements

This research was supported by Research Grant ES-00016, Office of Resource Development, Bureau of State Services, *U.S. Public Health Service*, Washington, D.C.

Table VI. *Chemical names of pesticides and related compounds mentioned in text*

Common name	Chemical name
Diazinon	0,0-diethyl-0-(2-isopropyl-4-methyl pyrimidinyl) phosphorothioate
Diazoxon	0,0-diethyl-0-(2-isopropyl-4-methyl pyrimidinyl) phosphate
Dipterex	dimethyl-trichloro-hydroxyethyl phosphonate
Methyl parathion	0,0-dimethyl 0-p-nitrophenyl phosphorothioate
Paraoxon	0,0-diethyl p-nitrophenyl phosphate
Parathion	0,0-diethyl-0-p-nitrophenyl phosphorothioate

Summary

Empirical observations note a comparatively long residual life of organophosphorous insecticides in aquatic environments under certain conditions. Hydrolysis rates of these compounds and their metabolites are of interest since chemical hydrolysis determines whether or not toxic residues will persist. Herein, the main objectives are evaluation of the kinetics of hydrolysis of diazinon and its oxygen analog diazoxon, and identification of major hydrolysis product(s). Hydrolysis of both compounds in the pH range 3.1 to 10.4 confirm a pseudo first-order rate expression. Rates are controlled by the $[H_3{}^+O]$, *i.e.*, acid or alkali catalysis. Diazinon hydrolysis was extremely rapid under acid conditions (pH 3.1, $t\frac{1}{2}$ = 706 minutes) whereas neutral or slightly alkaline solutions lower the rates ($t\frac{1}{2}$ at pH 7.4 and 9.0 are 4,435.8 and 3,263.0 hours, respectively). Raising the pH from 9.0 to 10.4 speeds the degradation process appreciably ($t\frac{1}{2}$ at pH 10.4 is 144.9 hours). Diazoxon hydrolysis in the pH range 3.1 to 10.4 proceeds much faster when compared to diazinon. Since pH values of natural waters range from 5.5 to 8.5, diazinon and diazoxon will have long residual lives. Identification of the major hydrolysis product 2-isopropyl-4-methyl-6-hydroxypyrimidine was by gas-liquid chromatography in combination with infrared and ultraviolet spectrophotometry.

Résumé *

Cinétique de l'hydrolyse du diazinon et diazoxon

Des observations empiriques indiquent une persistance résiduelle relativement longue des insecticides organo-phosphorés dans les milieux aquatiques, sous certaines conditions. Les taux d'hydrolyse de ces composés et de leurs métabolites présentent de l'intérêt, car l'hydrolyse chimique détermine si des résidus toxiques persisteront ou non. De ce fait, les principaux objectifs sont l'évaluation de la cinétique de l'hydrolyse du diazinon et de son homologue oxygéné, le diazoxon, ainsi que l'identification des principaux produits d'hydrolyse. L'hydrolyse des deux composés dans la gamme des pH de 3,1 à 10,4 confirme qu'il s'agit d'une pseudo-réaction du premier degré. Les vitesses d'hydrolyse sont contrôlées par la concentration en $[H_3{}^+O]$, c'est-à-dire la catalyse acide ou alcaline. L'hydrolyse du diazinon est extrêmement rapide en milieu acide (pH 3,1, t ½ = 706 minutes) ; plus lente en solution neutre ou légèrement alcaline (t ½ à pH 7,4 et 9,0 respectivement de 4.435,8 et 3.263,0 heures). Si l'on augmente le pH de 9,0 à 10,4, le processus de dégradation est notablement accéléré (t ½ à pH 10,4 est de 144,9 heures). L'hydrolyse du diazoxon dans la gamme des pH de 3,1 à 10,4 est beaucoup plus rapide, si on la com-

---

* Traduit par S. DORMAL-VAN DEN BRUEL.

pare à celle du diazinon. Puisque les valeurs de pH des eaux naturelles s'étendent de 5,5 à 8,5, le diazinon et le diazoxon auront une longue durée de persistance. L'identification du principal produit d'hydrolyse, le 2-isopropyl-4-méthyl-6-hydroxypyrimidine, a été effectuée par chromatographie gaz liquide combinée à la spectrophotométrie infra-rouge et ultra-violette.

Zusammenfassung *

Hydrolysekinetik von Diazinon und Diazoxon

Empirische Beobachtungen notieren eine verhältnismässig lange Rückstandslebenszeit von Organophosphorinsektiziden in Wasserumgebung unter bestimmten Bedingungen. Hydrolyseraten dieser Verbindungen und ihrer Metaboliten sind von Interesse, da chemische Hydrolyse dafür bestimmend ist, ob oder nicht toxische Rückstände beharren. Die Hauptaufgabe ist hier die Auswertung der Hydrolysekinetik von Diazinon und seines Oxydanalogen Diazoxon und die Identifizierung von Haupthydrolyseprodukten. Hydrolyse beider Verbindungen im pH-Bereich von 3.1 bis 10.4 bestätigen einen Ratenausdruck von Pseudo erster Ordnung. Raten sind kontrolliert durch (H_3O^+), das heisst, saure oder alkalische Katalyse. Diazinon Hydrolyse war äusserst schnell unter sauren Bedingungen (pH 3.1, t½ = 706 Minuten), während neutrale oder schwach alkalische Lösungen die Raten erniedrigen (t½ bei pH 7.4 und 9.0 sind 4435,8 bezw. 3263,0 Stunden). Beim Erhöhen des pH von 9.0 auf 10.4 wird der Abbauprozess bemerkenswert beschleunigt (t½ bei pH 10.4 ist 144 Stunden). Diazoxonhydrolyse im Bereich von 3.1 bis 10.4 geht viel schneller for sich im Vergleich zu Diazinon. Da pH-Werte natürlicher Wässer von 5.5 bis 8.5 rangieren, werden Diazinon und Diazoxon eine lange Rückstandslebenszeit haben. Identifizierung des Haupthydrolyseprodukts 2-isopropyl-4-methyl-6-hydroxypyrimidin wurde durch Gas-flüssig-Chromatographie in Verbindung mit Infrarot- und Ultraviolettspektrometrie vorgenommen.

References

Barnard, P. W. C., C. A. Bunton, D. R. Llewellyn, C. A. Vernon, and V. A. Welch: The reactions of organic phosphates. Part V. The hydrolysis of triphenyl and trimethyl phosphates. J. Chem. Soc., p. 2670 (1961).
Blumenthal, E., and J. B. Herbert: Hydrolysis of trimethyl orthophosphate. J. Trans. Faraday Soc. 41, 611 (1945).
Chanley, J. D., E. M. Gindler, and H. Sobotka: The mechanism of the hydrolysis of salicyl phosphate. J. Amer. Chem. Soc. 74, 4347 (1952).
Christian, G. D., and W. C. Purdy: The residual current in orthophosphate medium. J. Electroanalyt. Chem. 3, 363 (1962).

---

* Übersetzt von A. Schumann.

DOSTROVSKY, I., and M. HALMANN: Kinetic studies in the phosphinyl chloride and phosphorochloridate series. Part I. Solvolytic reactions. J. Chem. Soc., p. 502 (1953).

—— —— Kinetic studies in the phosphinyl chloride and phosphorochloridate series. Part V. Evidence for a one-stage mechanism in the hydrolysis of diethyl phosphorochloridate. J. Chem. Soc., p. 1004 (1956).

EL-REFAI, H. R., and L. GIUFFRIDA: Separation and micro-quantitative determination of Dipterex and DDVP by gas-liquid chromatography. J. Assoc. Official Agr. Chemists 48, 374 (1965).

FAUST, S. D., and I. H. SUFFET: Analysis of organic pesticides in aquatic environment. ASTM Symposium: Microorganics in water. In press (1969).

FUKUTO, T. R.: The chemistry and action of organic phosphorus insecticides. Adv. Pest Control Research V, 1 (1957).

GERSMAN, H. R., and J. A. KETELAAR: Chemical studies on insecticides. Rec. Travaux Chim. Pays-Bas 77, 1018 (1958).

GETZIN, L. W., and I. ROSEFIELD: Persistence of diazinon and zinophos in soils. J. Econ. Entomol. 59, 512 (1966).

GIUFFRIDA, L.: Isolation of microquantities of pesticides by gas chromatography for infrared. J. Assoc. Official Agr. Chemists 48, 354 (1965).

GYSIN, H., and A. MARGOT: Chemistry and toxicological properties of 0,0-diethyl-0-(2-isopropyl-4-methyl-6-pyrimidinyl) phosphorothioate (Diazinon). J. Agr. Food Chem. 6, 900 (1958).

HEATH, D. F.: The effect of substituents on the rates of hydrolysis of some organophosphorus compounds. Part I. Rates in alkaline solution. J. Chem. Soc., 3796 (1956 a).

—— The effect of substituents on the rates of hydrolysis of some organophosphorus compounds. Part II. Rates in neutral solution. J. Chem. Soc., p. 3804 (1956 b).

KANSOUTH, A. S. H., and T. L. HOPKINS: Diazinon absorption, translocation and metabolism in bean plants. J. Agr. Food Chem. 16, 446 (1968).

KETELAAR, J. A. A.: The hydrolysis of parathion and dimethyl-parathion. Rec. Trav. Chim. 69, 649 (1950).

KONRAD, J. G., D. E. ARMSTRONG, and G. CHESTERS: Soil degradation of diazinon, a phosphorothioate insecticide. Agron. J. 59, 591 (1967).

MORTLAND, M. M., and K. V. RAMAN: Catalytic hydrolysis of some organic phosphate pesticides by copper (II). J. Agr. Food Chem. 15, 163 (1967).

MUHLMANN, R., and G. SCHRADER: Hydrolyse der Insektiziden Phosphorsaureester. Z. Naturforsch. 12b, 196 (1957).

NICHOLSON, H. P., A. R. GRZENDA, G. J. LAUER, W. S. COX, and J. I. TEASLEY: Water pollution by insecticides in an agricultural river basin. I. Occurrence of insecticides in river and treated municipal water. Limnol. Oceanog. 9, 310 (1964).

O'BRIEN, R. D.: Toxic phosphorus esters—Chemistry, metabolism, and biological effects. New York: Academic Press (1960).

PECK, D. R.: Hydrolysis of parthion. Chem. & Ind. 67, 526 (1948).

PLIMMER, R., and BURCH, W.: Esters of phosphoric acid. Part I. Phosphates of cetyl alcohol, cholesterol, chloroethyl alcohol, and ethylene glycol. J. Chem. Soc., p. 502 (1953).

RALLS, J. W., D. R. GILMORE, and A. CORTES: Fate of radioactive 0,0-diethyl 0-(2-isopropyl-4-methylpyrimidin-6-yl) phosphorothioate on field-grown experimental crops. J. Agr. Food Chem. 14, 387 (1966).

ROSEN, A. A., and F. M. MIDDLETON: Chlorinated insecticides in surface waters. Anal. Chem. 31, 1729 (1959).

RUZICKA, J. H., J. THOMSON, and B. B. WHEALS: The gas chromatographic deter-

mination of organophosphorus pesticides. Part II. A comparative study of
hydrolysis rates. J. Chromatog. **31**, 37 (1967).

Suffet, I. H., S. D. Faust, and W. F. Carey: Gas-liquid chromatographic separation of some organophosphate pesticides, their hydrolysis products, and
oxons. Environ. Sci. and Technol. **I**, 639 (1967).

Swenson, H. A.: The montebello incident. Proc. Soc. Water Treatment **11**, 84
(1962).

Vernon, C. A.: The mechanisms of hydrolysis of organic phosphates. J. Chem.
Soc., Special Pub. No. 8, p. 17 (1957).

Weiss, C. M., and J. H. Gakstatter: The decay of anticholinesterase activity of
organic phosphorus insecticides on storage in waters of different pH. Proc.
2nd Internat. Water Pollution Research Conf. Tokyo, p. 83 (1964); New
York: Pergamon Press (1965).

Williams, E. F.: Properties of parathion and its phosphate analog. Ind. Eng.
Chem. **43**, 950 (1951).

Biodegradation kinetics of 2,4-dichlorophenoxyacetic acid by aquatic microorganisms *

By

ROLAND B. HEMMETT, JR.** AND SAMUEL D. FAUST **

Contents

* Paper of the Journal Series, New Jersey Agricultural Experiment Station, Rutgers, the State University of New Jersey, Department of Environmental Sciences, New Brunswick, N.J.
** Department of Environmental Sciences and Bureau of Conservation and Environmental Science, Rutgers, the State University, New Brunswick, N.J.

I. Introduction

Biological degradation of organic pesticides,[1] when provided as the sole substrate, has seldom been examined in aquatic systems. It would be extremely desirable, therefore, to develop a model system from which two pertinent questions could be answered: (a) What is the biological availability of various organic pesticides to degradation, and (b) What is the rate of this availability?

2,4-Dichlorophenoxyacetic acid (2,4-D), an aquatic and terrestrial herbicide, was selected as the reference or model compound for the following reasons: (a) it is manufactured, sold, and applied in large quantities each year, (b) its sodium salt is soluble in water which eliminates some experimental difficulties, and (c) a heterogeneous community of aquatic microorganisms can be developed or acclimatized to degrade it (Aly and Faust 1964).

The term "biodegradation" often may lead to confusion and misunderstanding. Biodegradation may be considered as the conversion of potentially harmful chemical compounds to innocuous substances by microorganisms. There are, however, a number of extents to which biodegradation may occur. "Primary" biodegradation is considered the minimum extent necessary to change molecular identity of the compound. "Acceptable" biodegradation is carried to the point at which some undesirable property is lost. "Ultimate" biodegradation is carried to water, carbon dioxide, and inorganic materials. The latter definition is assumed herein. Verification of the ultimate biodegradation of 2,4-D was obtained from the loss of light energy adsorption in the ultraviolet region and from oxygen uptake measurements.

II. Literature review

a) Biodegradation in soil systems

2,4-D has been shown to be biodegradable in soils by DeRose and Newman (1947), Brown and Mitchell (1948), and Audus (1949). Audus (1964) listed numerous microorganisms which are capable of degrading 2,4-D.

There have been many theories proposed concerning the nature of the enzyme system responsible for the biodegradation of 2,4-D. Audus (1952) proposed two theories. The first theory (mutation) assumed that upon the addition of 2,4-D to a microbial population, only one or two out of millions survived that would be able to degrade 2,4-D. The second theory (adaptive) assumed that a considerable proportion of the total population may be capable of responding to the herbicide as their responsive organisms grow and divide. Enzyme systems slowly

---

[1] Pesticides mentioned in text are identified in Table II.

become modified in the progeny to allow this foreign molecule to be used as a source of energy. AUDUS (1964) reported experiments which showed the latter theory (adaptation) to be correct.

NEWMAN and WALKER (1956) believed the mutation theory was correct. They also found that the apparent inability of cultures to grow well without 2,4-D points strongly towards a mutational origin.

STEENSON and WALKER (1956) stated that the ability to oxidize chlorophenoxyacetic acids depends on adaptive enzyme formation. Experiments with pure cultures of bacteria showed that the enzyme or enzymes were adaptive since they were possessed only by cells grown in the presence of a given herbicide or at least a close analogue. BELL (1957 and 1960) believed that it was improbable that bacteria in soil untreated with 2,4-D could possess enzymes specific for the decomposition of 2,4-D. It was possible, however, that cells with enzymes of low specificity, which could attack natural compounds structurally related to 2,4-D, could also attack 2,4-D without any qualitative change in their enzyme constitution. BELL stated this does not preclude the possibility that induced enzymes may be involved in the herbicide's degradation. The active cells might be mutants which were capable of forming induced enzymes for the decomposition of 2,4-D.

KEARNEY et al. (1966) stated that certain herbicides were degraded by "constitutive soil microorganisms" which require little or no adaptation prior to their active degradation of these chemicals. These authors also listed five criteria necessary for microbial decomposition of herbicides in the soil:

(1) "An organism which is effective in metabolizing the herbicide molecule must exist in the soil or must be capable of developing therein."

(2) "The compound must be in a form suitable for microbial degradation."

(3) "The chemical must be available to the organism."

(4) "The compound must be capable of inducing formation of the appropriate enzyme or enzymes concerned in the detoxication. Most enzymes concerned in these detoxication reactions require induction, few are constitutive. Low solubility, low concentration of the pesticide or permeability barriers within the organism may be associated with the lack of induction."

(5) "Environmental conditions such as soil pH, temperature and organic matter must be suitable for the microorganism to proliferate and for the enzyme to operate."

ALEXANDER and ALEEM (1961), reporting on the ease with which a microorganism may attack a herbicide molecule, found:

(1) "No compound having on its aromatic nucleus a chlorine in the meta position (that is at the 3 or 5 position), is transformed to an extent sufficient to cause a significant loss in ultraviolet absorption."

(2) "That the point of linkage of the fatty acid side chain (Alpha vs. Omega linkage) regulates the susceptibility of the aromatic portion of the molecule to microbial attack."

b) Metabolic pathways

The biological pathway for breakdown of 2,4-D has been studied extensively in soil systems. AUDUS (1952) believed the first step was an attack on the acetic acid side chain which was split off by hydrolysis, giving glycollic acid and leaving the corresponding phenol. EVANS and SMITH (1954), EVANS and MOSS (1957), and EVANS et al. (1961) reported the isolation from 2,4-D grown in pseudomonods cultures of a number of cleavage products. This suggested the sequential formation of 6-hydroxy-2,4-dichlorophenoxyacetate, 3,5-dichlorocatechol, and a chloromuconic acid.

STEENSON and WALKER (1957) found no oxygen uptake for 6-OH-2,4-D when applied to cells adapted to 2,4-D. These researchers argued from Staniers' simultaneous adaptation hypothesis which states that cells adapted to a compound should also be able to degrade any of the intermediate metabolites without a lag period. It was found that 6-OH-2,4-D could not be a metabolite because it was not degraded by cells adapted to metabolize 2,4-D. However, 2,4-dichlorophenol was proven a metabolite as was 4-chlorocatechol and possibly chlorocatechol.

BELL (1960) also stated that 2,4-dichlorophenol was a likely metabolite of 2,4-D. However, cells not adapted to 2,4-D could vigorously oxidize 2,4-dichlorophenol without an appreciable lag period. BELL believed the results showed these cells to possess constitutive enzymes. Since BELL found that 6-OH-2,4-D was oxidized by Achromobacter but not as well as 2,4-D, no intermediary role could be assigned to the compound until it was found to be oxidized only by 2,4-D adapted cells or preferably when it was isolated following the metabolism of 2,4-D. BELL stated that it is quite possible that each organism which can degrade phenoxyacetate does so via unique pathways, but the basic mechanisms involved may be much the same.

LOOS et al. (1967) found that although the 2,4-D metabolizing enzyme system of the Arthrobacter was inducible, a pathway for 2,4-D breakdown could not be established by the sequential induction technique alone. Compounds that could be considered as intermediates on the basis of sequential induction studies were 2-CPA, 4-CPA, 2-CP, 4-CP, 2,4-DP, 4-chlorocatechol, catechol, and possible 3,5-dichlorocatechol. LOOS pointed out that some of the compounds may not be intermediates in 2,4-D breakdown but rather may be metabolized because of their structural similarity to 2,4-D or 2,4-D-breakdown products.

c) Biodegradation in aquatic systems

Research into the persistence of 2,4-D in aquatic environments has been limited. OKEY and BOGAN (1962), using an activated sludge system, found that 2,4-D underwent considerable metabolism. They found, however, that cultures which were adapted to degrade 2,4-D could not degrade 2,4,5-T.

FAUST and ALY (1964) used bottom mud samples collected from two lakes. The first lake had been treated the previous year with 2,4-D while the second lake had never been treated with 2,4-D. It was observed that when 2,4-D was applied to the mud from the lake previously treated with 2,4-D, it disappeared in 35 days, whereas 2,4-D persisted in muds from the untreated lake for 65 days. These authors emphasized the importance of the times involved in the degradation (35 and 65 days) of the two samples. These times when compared with soil samples, which require about two weeks, show that the biodegradation in lakes is much slower.

SCHWARTZ (1967) found that only a small fraction of 2,4-D was degraded by a mixed microbial population in a dilute medium of mineral salts. The presence of larger amounts of nutrient broth as a supplemental source of organic carbon had no appreciable effect on the rate of decomposition.

DEMARCO et al. (1967) investigated several environmental conditions under which 2,4-D would persist or be biodegraded in natural lake waters. In "warm" aerobic waters, 2,4-D was biodegraded, apparently, within six days whereas 2,4-D persisted for as long as 80 days in "cold" deoxygenated waters. These "warm" and "cold" conditions were selected to simulate stratified and impounded reservoir waters.

It becomes apparent from the above literature review that 2,4-D is degraded quite readily in terrestrial environments whereas conditions in aquatic environments may not be as favorable for rapid disappearance. It also appears that the kinetics of 2,4-D biodegradation have not been examined in an extensive manner.

III. Materials and methods

a) Enrichment cultures

Domestic activated sewage sludge was used as a source of microorganisms. The procedure of HUNTER and HEUKELEKIAN (1964) was employed to acclimatize this heterogeneous population of aquatic microorganisms to 2,4-D. This technique involves the sequential addition of increasing amounts of 2,4-D to the microorganisms over a period of 12 to 14 days. The acclimatized microorganisms were maintained by periodic additions of the 2,4-D and a combination of 2,4-D plus synthetic sewage. An aerobic environment was maintained by a continuous shaking device.

b) Manometric techniques

Determination of the mass of acclimatized cells was obtained by measuring the total dry solids concentration (*Standard Methods* 1965). A specific quantity of acclimatized cells was added to 100-ml. volumetric flasks. The following nutrients then were added to the flasks: $MgSO_4$, $FeCl_3$, $CaCl_2$, and PO_4 buffer. To the control flasks, distilled water was added to bring the volume to 100-ml. To the enrichment flasks, the desired amount of 2,4-D was added, with the volume brought to 100-ml. by distilled water.

Aliquots (50-ml.) from the control and enrichment flasks then were placed in separate Warburg flasks. In the center well of each flask was placed 1.0-ml. of potassium hydroxide. Oxygen uptake subsequently was measured for the appropriate period of time. All tests were conducted at $20° \pm 0.5°$ C. The pH was maintained at 7.0 ± 0.2.

c) Analytical methods

2,4-D residuals were determined at 284 mμ (ALY and FAUST 1963). This analytic information confirmed disappearance of the 2,4-D molecule in the aquatic microorganism system.

d) Materials

Eastman Kodak, white label grade, 2,4-dichlorophenoxy acetic acid was used.

IV. Experimental results

a) Kinetic model

The biological degradation of 2,4-D was assumed to follow the stoichiometry suggested by this reaction:

$$+ 7.5O_2 \xrightarrow[\text{[A.M.]} = K]{} 8CO_2 + 2H_2O + 2HCl \qquad (1)$$

Thus 1.00 mg. of 2,4-D should require 1.09 mg. of oxygen for complete conversion to carbon dioxide. In an examination of the biodegradation kinetics of equation (1), three variables must be considered.

Many substrates disappear in biological systems in accord with zero-order kinetics:

$$\frac{-d\,[2,4\text{-}D]}{dt} = k\,[2,4\text{-}D]^\circ \tag{2}$$

Thus, rate of disappearance of 2,4-D is independent of initial 2,4-D concentration. The integrated form of equation (2) is:

$$[2,4\text{-}D]_t = -kt + [2,4\text{-}D]_{to} \tag{3}$$

where $[2,4\text{-}D]_{to}$ = initial concentration in mg./l., $[2,4\text{-}D]_t$ = concentration at time t in mg./l., and k = rate constant in mg. $l.^{-1}t^{-1}$. The k was evaluated from a plot of $[2,4\text{-}D]_t$ *versus* time.

Equations (1), (2), and (3) hold the microorganism concentration constant. It can be observed, however, that the rate constant, k, is a function of the microorganism concentration. That is, as the latter is increased, the k value increases in accord with the equation:

$$\frac{-d\,[2,4\text{-}D]}{dt} = k'\,[A.M.] = k\,[2,4\text{-}D]^\circ \tag{4}$$

where [A.M.] = concentration of acclimatized microorganisms and k' = proportionality constant. Rearranging equation (3) yields:

$$k = \frac{k'\,[AM]}{[2,4\text{-}D]^\circ} \tag{5}$$

Hence an increase in concentration of microorganisms increases the rate of utilization of 2,4-D. Microorganism concentration is expressed herein as mg./l. dry weight.

The rate constant, k, was evaluated from a linear plot of 2,4-D concentration *versus* time in accord with equation (3). There was observed, however, a lag phase in oxygen uptake whenever the acclimatized microorganisms were transferred from the enrichment flasks into the Warburg flasks. Herein the k values were taken from the linear portion of the plots after a lag phase of four to five hours.

b) Herbicide concentration held constant, microorganism concentration varied

Figure 1 shows that the rate of oxidation does increase with an increase in the microorganism concentration. The k values are 1.20 and 2.10 mg. $l.^{-1}t^{-1}$ for microorganism concentrations of 500 and 1,000 mg./l., respectively. Figure 2 shows the same thing except that a herbicide concentration of 100 mg./l. was used. The k values for Figure 2 are 1.89 and 4.11 mg. $l.^{-1}t^{-1}$ for microorganism concentrations of 500 and 1,000 mg./l., respectively. It is quite apparent that there is a definite increase in the rate constant of oxidation with an increase in the microorganism concentration.

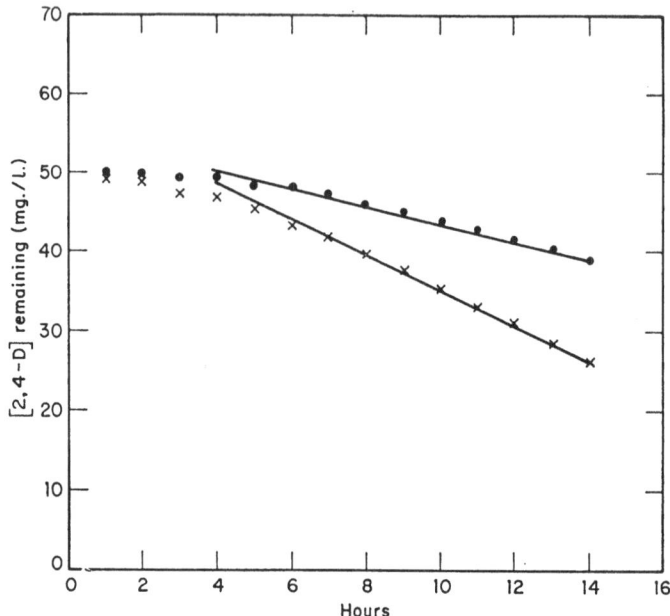

Fig. 1. The biodegradation rates of 50 mg./l. of 2,4-D using two different concentrations of microorganisms: ●——●, [A.M.] = 500 mg./l.; ×——×, [A.M.] = 1,000 mg./l.

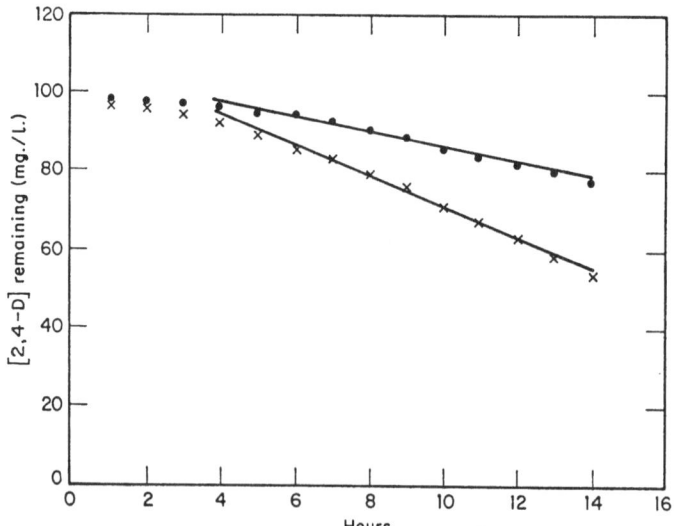

Fig. 2. The biodegredation rates of 100 mg./l. of 2,4-D using two different concentrations of microorganism: ●——●, [A.M.] = 500 mg./l.; ×——×, [A.M.] = 1,000 mg./l.

c) *Microorganism concentration held constant, herbicide concentration varied*

Figure 3 shows that as the concentration of the herbicide increases,

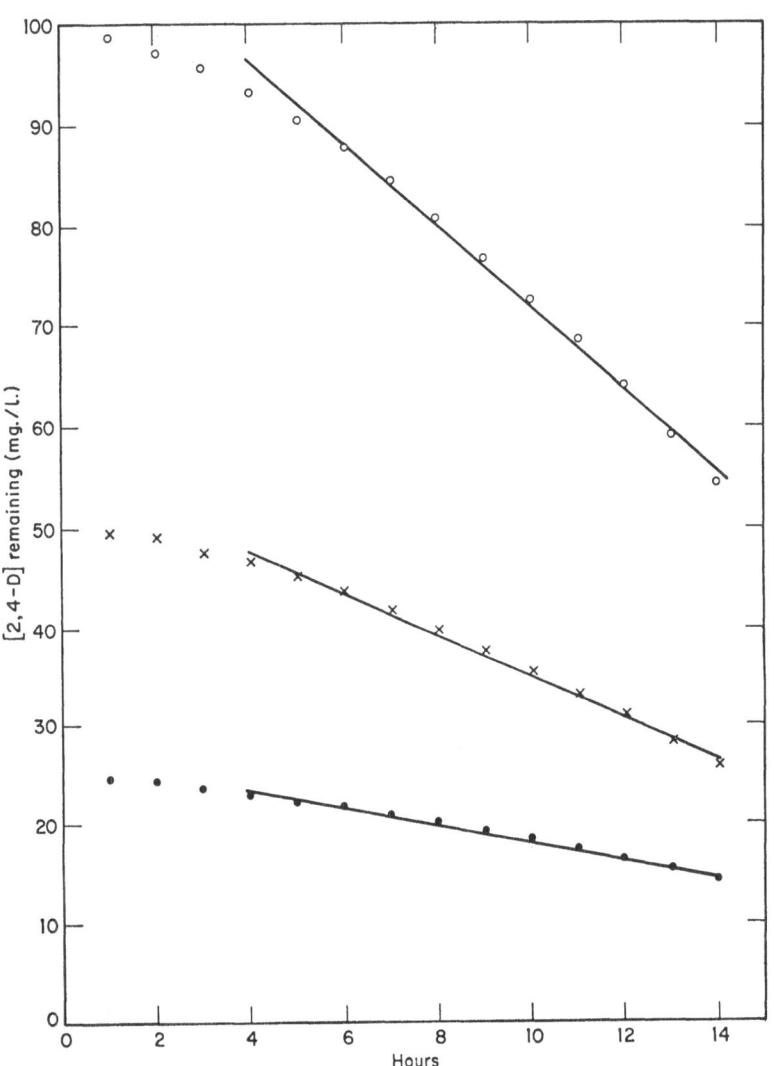

Fig. 3. The biodegradation rates of 2,4-D using a constant microorganism concentration (1,000 mg./l.) and three different herbicide levels: O——O, [2,4-D] = 100 mg./l.; ×——×, [2,4-D] = 50 mg./l.; and ●——●, [2,4-D] = 25 mg./l.

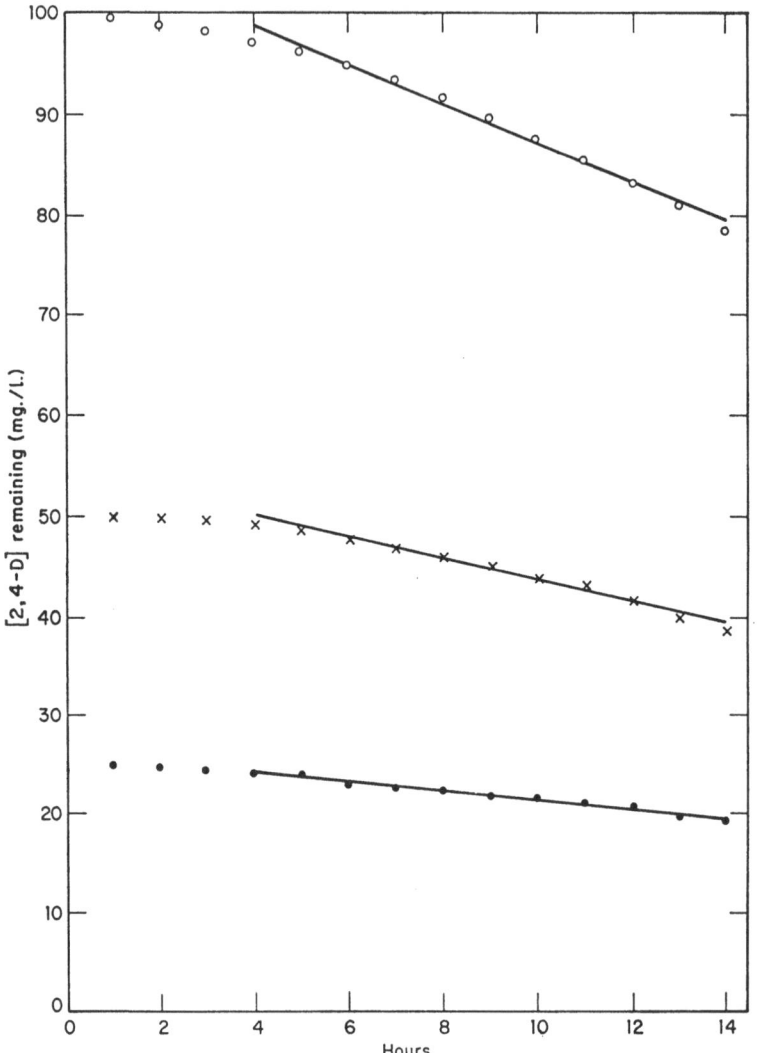

Fig. 4. Biodegradation rates of 2,4-D using a constant microorganism concentration (500 mg./l.) and three different herbicide levels: O——O, [2,4-D] = 100 mg./l.; ×——×, [2,4-D] = 50 mg./l.; and ●——●, [2,4-D] = 25 mg./l.

the rate of oxidation also increases. The k values for Figure 3 are 0.885, 2.24, and 3.98 mg. l.$^{-1}$t^{-1} for herbicide concentrations of 25, 50, and 100 mg./l., respectively. Figure 4 indicates the same trend with k values of 0.47, 1.08, and 1.89 mg. l.$^{-1}$t^{-1} for herbicide concentrations of 25, 50, and 100 mg./l., respectively. It would appear that the rate constant of oxidation is dependent also on the concentration of the herbicide.

Confirmation of zero-order kinetics, in this case, is to see if the percent of 2,4-D oxidized/unit time is the same. Using Figure 3 as an example, the percent of 2,4-D oxidized between the ninth and tenth hours was 4.0, 4.0, and 4.5 for herbicide concentrations of 25, 50, and 100 mg./l., respectively. Therefore, the oxidization rate was independent of initial 2,4-D concentration.

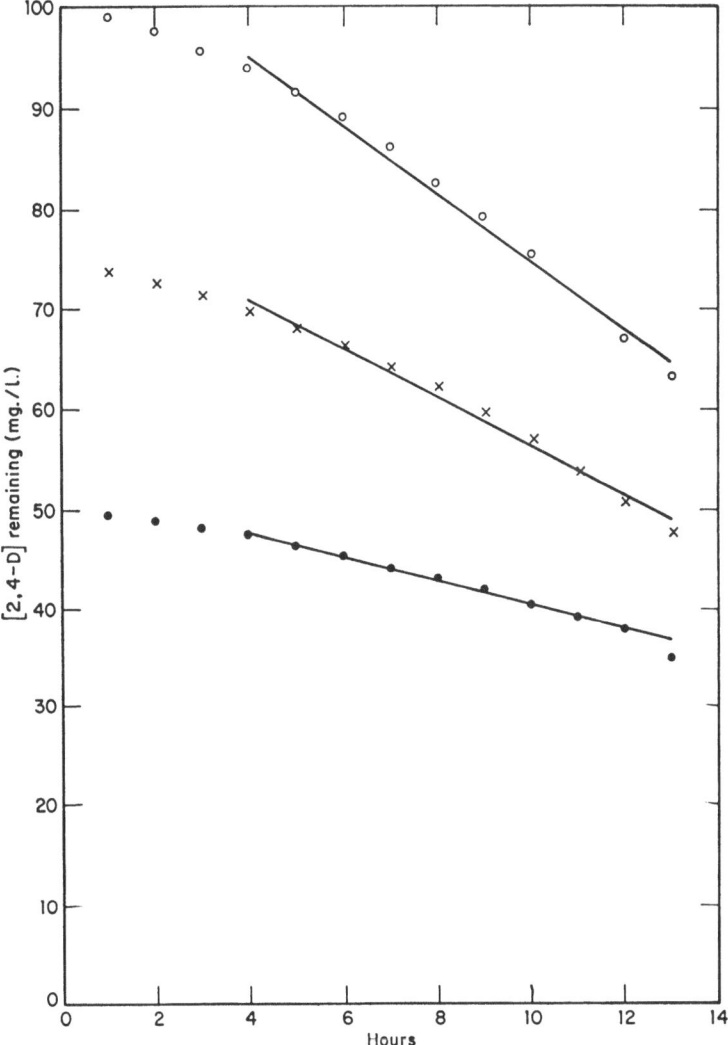

Fig. 5. The biodegradation rates of 2,4-D with the concentration of microorganisms to herbicide held at a ratio of 10:1 by weight: ○——○, [A.M.] = 1,000 mg./l.; ×——×, [A.M.] = 750 mg./l.; and ●——●, [A.M.] = 500mg./l.

d) Microorganism and herbicide concentrations varied with their ratios held constant

In Figure 5, the A.M. and 2,4-D were varied but their weight ratios were held constant. Apparent zero-order kinetics were observed again since the percentages of oxidation/hour were 3.0, 3.3, and 3.5 for herbicide concentrations of 50, 75, and 100 mg./l., respectively. The rate constants were 1.25, 2.42, and 3.37 mg. l.$^{-1}$t^{-1}, respectively, and are summarized in Table I.

Table I. *Summary of zero-order rate constants for the biodegradation of 2,4-D by aquatic microorganisms*

[A.M.] (mg./l. dry weight)	k value (mg. l.$^{-1}$t^{-1})			
	[2,4-D] (mg./l.)			
	25	50	75	100
500 [a]	—	1.20	—	2.10
1,000	—	1.68	—	4.11
500 [b]	.47	1.08	—	1.89
1,000	.885	2.24	—	3.98
500 [c]	—	1.25	—	—
750	—	—	2.42	—
1,000	—	—	—	3.37

[a] [2,4-D] held constant, [A.M.] varied.
[b] [A.M.] held constant, [2,4-D] varied.
[c] Ratio of [A.M.] to [2,4-D] held constant at 10:1.

e) Effect of acclimatization age on rate of oxidation

The length of time that an enrichment culture is exposed to a substrate affects its subsequent rate of oxidation. Figure 6 suggests that the longer the acclimatization period the shorter is the time for "complete" oxidization of 2,4-D. "Complete" oxidization was observed when 68 percent of the stoichiometric quantity of oxygen in equation (1) was recorded. There is, however, an apparent limitation to the acclimatization period as witnessed by the leveling of the curve in Figure 6. All cultures used herein were taken from the enrichment flasks after 60 days.

V. Discussion

The results suggest that the biodegradation of 2,4-D follows zero-order kinetics. The most important constituent of the reaction is, therefore, the size and nature of the microbial population. It has been observed that approximately 68 percent of the 2,4-D concentration is

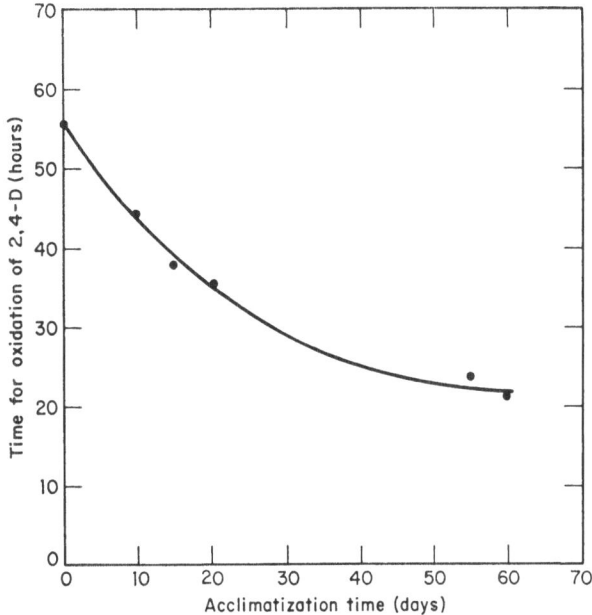

Fig. 6. The effect of acclimatization age on the rate of biodegradation

used in the respiration process as measured by the manometric techniques whereas 32 percent is incorporated into the microbial cells as a reserve source of energy.

The rates at which the biodegradation is performed are dependent on several variables. The first is the time necessary for the enzyme systems to become acclimatized to the herbicide. The second is the effect that the age of the acclimatized microorganisms has on the rate. It has been shown that the rate decreases as the acclimatization period increases until it reaches a stabilization point. Therefore, each new addition of 2,4-D should be degraded faster than the previous one provided the stabilization point has not been reached. The third variable is the natural condition of the aquatic environment. ALY and FAUST (1964) found that it took 65 days for 2,4-D to disappear from the bottom mud of a lake while it takes only 12 to 14 days for the same phenomenon to occur in the soil. The conditions may, therefore, be more severe in the aquatic environment concerning the biodegradation of herbicides. SCHWARTZ (1967) stated his studies showed that metabolism of 2,4-D was much slower than in the terrestrial environment. This could mean that pesticides which take considerably longer to degrade would persist for extremely long periods in the aquatic environment, thus increasing the chances of pollution to occur. Some support for this was reported by DEMARCO et al. (1967). An initial period of 12 to 14 days was utilized herein for the acclimatization of aquatic microorganisms to biodegrade 2,4-D. These microorganisms had not been exposed previously to 2,4-D and related compounds. On the other hand, if the acclimatization period is extended beyond 14 days (refer to Fig. 6), the rate at which 2,4-D is biodegraded steadily increases to a limiting or maximum value for a given concentration of microorganisms. This limitation is due, apparently, to a maximum enzyme concentration. If so, this is in accord with the well-known concepts of enzyme kinetics.

That substrate concentration is not a limiting factor in the kinetics of 2,4-D biodegradation was seen in Figure 5. Here the ratio of A.M. to 2,4-D was held at 10:1 for three concentrations of 2,4-D. When the kinetics were expressed as a percentage oxidation/unit time, relatively constant rates were observed in each of the three systems. This suggests an enzyme-saturated condition was in existence and adds substantiation to the zero-order kinetic model. Similar kinetic observations were reported by Stumm-Zollinger (1966) in glucose-aquatic microorganism systems.

Acknowledgments

This research was sponsored by Research Grant ES-00016, Office of Resource Development, Bureau of State Services, U.S. *Public Health Service, Washington, D.C.*

Table II. *Chemical names of pesticides mentioned in text*

Common name	Chemical name
2-CP	2-Chlorophenol
4-CP	4-Chlorophenol
2,4-DP	2,4-Dichlorophenol
2-CPA	2-Chlorophenoxyacetic acid
4-CPA	4-Chlorophenoxyacetic acid
2,4-D	2,4-Dichlorophenoxyacetic acid
2,4,5-T	2,4,5-Trichlorophenoxyacetic acid

Summary

The biological degradation of 2,4-dichlorophenoxyacetic acid (2,4-D) was assumed to follow the stoichiometry suggested by the following equation:

$$+ 7.5O_2 \longrightarrow 8CO_2 + 2H_2O + 2HCl$$

In a study of the biodegradation kinetics of 2,4-D three variables must be considered: (a) microorganism concentration, (b) substrate concentration, and (c) ratio of microorganism concentration to substrate concentration. These three variables affect the rate at which 2,4-D is converted to carbon dioxide, water, and hydrochloric acid.

The biodegradation appeared to follow zero-order kinetics, with the oxidation rate independent of the substrate concentration. When the ratio of microorganism concentration to 2,4-D concentration was held constant, constant rates of oxidation were observed which suggested the enzyme systems were saturated with substrate. Thus, substrate concentration was not a limiting factor in these kinetic studies, further substantiating the zero-order kinetic model. Acclimatization time also appeared as a factor. The longer the period of acclimatization, the shorter was the period of time to effect "ultimate" biodegradation of 2,4-D.

Résumé *

Cinétiques de biodégradation de l'acide 2,4-dichlorophenoxy acétique par les microorganismes aquatiques

La stoechiométrie de la dégradation biologique de l'acide 2,4-dichlorophenoxy acétique (2,4-D) a été, par hypothèse, définie par l'équation suivante :

$$+ 7.5O_2 \longrightarrow 8CO_2 + 2H_2O + 2HCl$$

Pour l'étude des cinétiques de biodégradation du 2,4-D, trois variables diovent être considérées : (a) la concentration du microorganisme, (b) la concentration du substrat et (c) le rapport de la concentration du microorganisme à celle du substrat. Ces trois variables affectent la vitesse avec laquelle le 2,4-D est transformé en gaz carbonique, eau et acide chlorhydrique. La biodégradation a paru suivre une réaction d'ordre zéro où la vitesse d'oxydation est indépendante de la concentration du substrat. Lorsque le rapport de la concentration du microorganisme à celle du 2,4-D était maintenu constant, des vitesses d'oxydation constantes étaient observées , signe de la saturation par le substrat des systèmes enzymatiques. Ainsi, la concentration du substrat n'était pas un facteur limitatif dans ces études cinétiques, confirmant

---

* Traduit par R. MESTRES.

le modèle cinétique d'ordre zéro. Un temps d'acclimatation apparut aussi comme facteur. Plus la période d'acclimatation est longue, plus est courte celle nécessaire pour effectuer l'ultime dégradation du 2,4-D.

Zusammenfassung *

Bioabbaukinetik von 2,4-Dichlorphenoxyessigsäure durch Wassermikroorganismen

Es wurde angenommen, dass der biologische Abbau von 2,4-Dichlorphenoxyessigsäure (2,4-D) der Stöchiometrie folgt, die durch die folgende Gleichung angegeben ist:

$$\text{(2,4-Dichlorphenoxyessigsäure)} + 7.5 O_2 \longrightarrow 8 CO_2 + 2 H_2O + 2 HCl$$

In einer Studie der Bioabbaukinetik von 2,4-D müssen 3 veränderliche Grössen in Erwägung gezogen werden: (a) Mikroorganismenkonzentration, (b) Substratkonzentration und (c) das Verhältnis von Mikroorganismenkonzentration zu Substratkonzentration. Diese 3 veränderlichen Grössen beeinflussen die Rate, mit welcher 2,4-D in Kohlendioxid, Wasser und Salzsäure umgewandelt wird.

Der Bioabbau scheint der Null-Ordnung-Kinetik zu folgen, wobei die Oxidationsrate unabhängig von der Substratkonzentration ist. Wenn das Verhältnis von Mikroorganismenkonzentration zu 2,4-D-konzentration konstant gehalten wurde, wurden konstante Oxidationsraten beobachtet; dies legt nahe, dass die Enzymsysteme mit Substrat gesättigt waren. Daher war die Substratkonzentration nicht der begrenzende Faktor in diesen kinetischen Studien, was weiterhin das Null-Ordnungs-kinetische Modell bestätigt. Auch die Eingewöhnungszeit erschien als ein Faktor. Je länger die Zeit der Eingewöhnung, je kürzer war die Zeit, um letztlichen Bioabbau von 2,4-D zu bewirken.

References

ALEXANDER, M., and M. I. H. ALEEM: Effect of chemical structure on microbial decomposition of aromatic herbicides. J. Agr. Food Chem. 9, 44 (1961).

ALY, O. M., and S. D. FAUST: Determination of 2,4-dichlorophenoxy acetic acid in surface waters. J. Amer. Water Works Assoc. 55, 639 (1963).

—— —— Studies on the fate of 2,4-D and ester derivatives in natural surface waters. J. Agr. Food Chem. 12, 541 (1964).

---

* Übersetzt von A. SCHUMANN.

AUDUS, L. J.: Biological detoxication of 2,4-dichlorophenoxy acetic acid. Plant and soil **2**, 31 (1949).
—— Decomposition of 2,4-dichlorophenoxyacetic acid and 2-methyl-4-chlorophenoxyacetic acid in the soil. J. Sci. Food Agr. **3**, 268 (1952).
—— Herbicide behavior in the soil. In: The physiology and biochemistry of herbicides. New York-London: Academic Press (1964).
BELL, G. R.: Some morphological and biochemical characteristics of a soil bacterium which decomposes 2,4-dichlorophenoxyacetic acid. Can. J. Microbiol. **3**, 821 (1957).
—— Studies on a soil Achromobacter species which degrades 2,4-dichlorophenoxyacetic acid. Can. J. Microbiol. **6**, 1325 (1960).
BROWN, J. W., and J. W. MITCHELL: Inactivation of 2,4-dichlorophenoxyacetic acid in soil as affected by soil moisture, temperature and the addition of manure and autoclaving. Botan. Gaz. **109**, 314 (1948).
DEMARCO, J., J. M. SYMONS, and G. G. ROBECK: Behavior of synthetic organics in stratified impoundments. J. Amer. Water Works Assoc. **59**, 965 (1967).
DEROSE, H. R., and A. S. NEWMAN: The comparison of the persistence of certain plant growth regulators when applied to soil. Proc. Soil Sci. Soc. Amer. **12**, 222 (1947).
EVANS, W. C., and P. MOSS: The metabolism of the herbicide p-chlorophenoxyacetic acid by a soil micro-organism. Biochem. J. **65**, 8P (1957).
——, and B. S. W. SMITH: The photochemical inactivation and microbial metabolism of the chlorophenoxyacetic acid herbicides. Biochem. J. **57**, xxx (1954).
——, J. K. GAUNT and J. I. DAVIES: The metabolism of chlorophenoxyacetic acid herbicides by soil micro-organisms. Congr. Internat. Biochem. **5**, 306 (1961).
HUNTER, J. V., and H. H. HEUKELEKIAN: Determination of biodegradability using Warburg respirometric methods. Proc. 19th Indust. Waste Conf., Purdue Univ., p. 616 (1964).
KEARNEY, P. C., D. O. KAUFMAN, and M. ALEXANDER: Biochemistry of herbicides decomposition in soil. Symp. organic pesticides in the environment. Amer. Chem. Soc. (1966).
LOOS, M. A., R. N. ROBERTS, and M. ALEXANDER: Phenols as intermediates in the decomposition of phenoxyacetates by an Arthrobacter species. Can. J. Microbiol. **13**, 679 (1967).
NEWMAN, A., and R. L. WALKER: Microbiological decomposition of 2,4-dichlorophenoxyacetic acid. Applied Microbiol. **4**, 201 (1956).
OKEY, R. W., and R. A. BOGAN: Apparent involvement of electronic mechanism in limiting the microbial metabolism of pesticides. J. Water Pollut. Control Fed. **37**, 692 (1965).
SCHWARTZ, H. J.: Microbial degradation of pesticides in aqueous solutions. J. Water Pollut. Control. Fed. **39**, 1701 (1967).
Standard Methods: Examination of water, sewage, and industrial wastes, 12 ed. Amer. Public Health Assoc., N.Y. (1965).
STEENSON, T. I., and N. WALKER: Bacterial oxidation of chlorophenoxyacetic acids. Plant and Soil **8**, 17 (1956).
STEENSON, T. I., and N. WALKER: The pathway of breakdown of 2,4-dichlorophenoxyacetic acid and 4-chloro-2-methyl-phenoxyacetic acid by bacteria. J. Gen. Microbiol. **16**, 146 (1957).
STUMM-ZOLLINGER, E.: Effects of inhibition and repression on the utilization of substrates by heterogeneous bacterial communities. Applied Microbiol. **14**, 654 (1966).

Subject Index

Acrolein 110
Aldrin 57, 63
―――― persistence in soils 138, 139
Alfalfa 15, 57
―――― hay and products, decontamination 15 ff.
―――― meal 14
―――― seed 121
Almond hull meal, decontamination 16 ff.
―――― hulls and meal 13, 14
Ametryne 115
Amiben 115
―――― persistence in soils 139
Amitrole 121, 126, 127
Animal feeds, decontamination 13 ff., 57 ff.
―――― feeds, definition 14
―――― feeds, ingredients 29 ff.
―――― feeds, residues in (see also specific pesticides) 2, 13 ff.
―――― feeds, tolerances for pesticides in 14
―――― products, reducing residues in 51 ff.
Apple pulp 14
Apples 40, 63, 69
Ashing of pesticides 91, 96 ff.
Atrazine 110, 115, 122, 124
―――― deliberate destruction 90, 91, 93, 95–98, 100
―――― persistence in soils 139, 146
Avocado leaves 41
Azinphos methyl 63, 69
―――― methyl, removal from oranges 40

Barban 125
―――― persistence in soils 139
Beans 4, 40, 42, 69, 74, 75, 78–82, 126, 185
Bensulide 119, 120
―――― persistence in soils 139
BHC, persistence in soils 139
Bidrin 6
Biodegradation, definition 192
Blanching to reduce residues 69, 78 ff.
Broccoli 42, 74–78, 80, 82
Bromacil 110, 115
―――― deliberate destruction 90, 91, 93, 95–101

―――― photodecomposition 141, 142
Butter 65

Canning operations and residues 69, 74 ff.
―――― operations to reduce residues 73 ff.
―――― wastes, pesticides in 82 ff.
Captan degradation 5, 9
Carbaryl and canning operations 75 ff.
―――― deliberate destruction 90, 91, 93–95, 97–101
―――― removal 69, 76, 78–81, 83
Carbon, see also Charcoal
―――― adsorption isotherms 155 ff.
―――― interaction with diquat and paraquat 151 ff.
Carrots 63
―――― as soil residue scavengers (see also Trap plants) 80
Catch plants, see Trap plants
CDAA, persistence in soils 139
CDEC 115
―――― persistence in soils 139
Celery, reduction of parathion residues on 39 ff.
Charcoal, see also Carbon
―――― to adsorb herbicides 117
―――― to mobilize DDT in fat storage 56 ff.
Cheese making, effects on residues 65
Chemical destruction of pesticides (see also Pesticides) 100 ff.
Chickens (see also specific pesticides) 64
Chlordane 57, 68
―――― in milk 65, 66
―――― persistence in soils 138, 139
Chloropicrin 108
Chloroxuron 120
CIPC 114
―――― persistence in soils 139
Clay mineral adsorption isotherms 156 ff.
Clays, interaction with diquat and paraquat 151 ff.
Clover seed screenings, decontamination 16 ff.
Coconut oil 55, 56
Combustion of pesticides 91, 96, 97
Composting and effect on residues 75 ff., 83

209

Manuscripts in Press

Distribution of pesticides in immiscible binary solvent systems for cleanup and identification and its application in the extraction of pesticides from milk. By Morton Beroza, May N. Inscoe, and Malcolm C. Bowman.

Factors affecting the extraction of organochlorine insecticides from soil. By M. Chiba.

Pesticide regulations and residue problems in Poland. By T. Stobiecki.

Photochemistry of halogenated herbicides. By J. R. Plimmer.

Leaf structure as related to absorption of pesticides and other compounds. By Herbert M. Hull.

Special volume: Triazine-soil interactions.

> History of the development of triazine herbicides. By E. Knüsli.
>
> Review of the use and performance of triazine herbicides on major crops and major weeds throughout the world. By A. Gast
>
> Introduction to triazine-soil interactions. By P. Dubach.
>
> Theory of adsorption and leaching of pesticides in soils, including a short review on soil chemistry. By J. L. White and G. W. Bailey.
>
> Adsorption of triazine herbicides on clay minerals. By J. B. Weber.
>
> Adsorption of triazine herbicides on soil organic matter, including a survey on methods used for such studies. By Charles S. Helling.
>
> Review of the influence of triazine herbicides on the soil microflora. By P. Kaiser and J. Pochon.
>
> Microbial degradation of triazine herbicides. By D. D. Kaufman and P. C. Kearney.
>
> Volatilization and non-biological degradation of triazine herbicides in vitro and in soils. By L. S. Jordan, W. J. Farmer, J. R. Goodin, and B. E. Day.
>
> Persistence of triazine herbicides and related problems. By T. J. Sheets.
>
> Ways and means to influence the activity and the persistence of triazine herbicides in soils. By H. M. LeBaron.
>
> Quantitative determination of triazine herbicides in soils by bioassay. By R. Behrens.
>
> Quantitative determination of triazine herbicides in soils by chemical analysis. By A. M. Mattson, R. A. Kahrs, and R. T. Murphy.
>
> Summary and conclusions. By P. C. Kearney.

Special volume: The chemistry of pesticides. By N. N. Mel'nikov. Translated from the Russian by Ruth L. Busbey.